Essential
Computer Hardware

Second Edition

Kevin Wilson

Elluminet Press
www.elluminetpress.com

Essential Computer Hardware: 2nd Ed

Publisher: Elluminet Press
Director: Kevin Wilson
Lead Editor: Steven Ashmore
Technical Reviewer: Mike Taylor, Robert Ashcroft
Copy Editors: Joanne Taylor, James Marsh
Proof Reader: Robert Ashcroft
Indexer: James Marsh
Cover Designer: Kevin Wilson

eBook versions and licenses are also available for most titles. Any supplementary material referenced by the author in this text is available to readers at

www.elluminetpress.com/resources

Table of Contents

About the Author

With over 15 years' experience in the computer industry, Kevin Wilson has made a career out of technology and showing others how to use it. After earning a master's degree in computer science, software engineering, and multimedia systems, Kevin has held various positions in the IT industry including graphic & web design, building & managing corporate networks and computer systems, as well as training and IT support.

He currently serves as Elluminet Press Ltd's senior writer and director, he periodically teaches computer science at college in South Africa and serves as an IT trainer in England. His books have become a valuable resource among the students in England, South Africa and our partners in the United States.

Kevin's motto is clear: "If you can't explain something simply, then you haven't understood it well enough." To that end, he has created the Exploring Technology Series, in which he breaks down complex technological subjects into smaller, easy-to-follow steps that students and ordinary computer users can put into practice.

Acknowledgements

Thanks to all the staff at Luminescent Media & Elluminet Press for their passion, dedication and hard work in the preparation and production of this book.

To all my friends and family for their continued support and encouragement in all my writing projects.

To all my colleagues, students and testers who took the time to test procedures and offer feedback on the book

Finally thanks to you the reader for choosing this book. I hope it helps you gain a better understanding of computer systems.

Computer Fundamentals

A computer is a machine that can store and process data according to a sequence of instructions called a program.

At their most fundamental level, these machines can only process binary data: 1s and 0s.

In this chapter, we'll take a look at using the binary code to encode data, as well as binary arithmetic and number bases.

We'll look at using logic gates to build simple circuits and how they form the building blocks for electronic devices, before moving onto the fetch execute cycle and instruction sets.

Let's begin by taking a look at the binary code.

Representing Data

The computer uses 1s and 0s to encode computer instructions and data. RAM is essentially a bank of switches: 'off' represents a 0 and 'on' represents a 1.

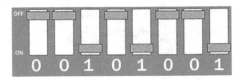

Using this idea, data can be encoded using either ASCII or Unicode and stored in RAM or on a disc drive.

ASCII code

The American Standard Code for Information Interchange (ASCII), originally used a 7-bit binary code to represent letters, numbers and other characters. Each character is assigned a binary number between 0 to 127. For example:

Capital A is 01000001_2 (65_{10})
Lowercase a is 01100001_2 (97_{10})

000-31 is reserved for control characters such as end of line, carriage returns, end of text, and so on.

032-126 covers symbols, numbers 0-9, and all lowercase and uppercase letters.

The ASCII code set was later extended to 8-bit which allowed more characters to be encoded. These included mathematical symbols, international characters and other special characters needed.

Unicode

Unicode is a universal encoding standard for representing the characters of all the languages of the world, including those with larger character sets such as Chinese, Japanese, and Korean.

UTF-8 is a variable length encoding system that uses one to four bytes to represent a character. UTF-8 is backwards compatible with ASCII and widely used in internet web pages. In your HTML code you might see something like this: <meta charset="utf-8">.

Binary Numbers

Binary numbers only have ones and zeros, in other words you can only represent a number using a 1 and a 0. We do this using the binary or base 2 number system. Using this system we can count in binary. Here's a table to help you:

Decimal	Binary
1	0001
2	0010
3	0011
4	0100
5	0101
6	0110
7	0111
8	1000
9	1001
10	1010

With decimal numbers (or base 10), if you remember from primary school mathematics, reading from right to left you have your ones, tens, hundreds, thousands and so on. You get these by writing out your powers of 10.

$10^0 = 1$

$10^1 = 10$

$10^2 = 100$

$10^3 = 1000$

So to write the number 123, you'd have 3 'ones', 2 'tens', and 1 'hundred'. Adding these together you get 123.

100	10	1		
1	2	3	=	One 100 + Two 10s + Three 1s = 123

With binary it's the same principle, except you use 2s instead of 10s. You get these by writing out your powers of 2.

$2^0 = 1$

$2^1 = 2$

$2^2 = 4$

$2^3 = 8$

So to read the number $1\ 1\ 1\ 1\ 0\ 1\ 1_2$, reading from right to left you'd have 1 'ones', 1 'twos', 0 'fours', 1 'eights', 1 'sixteens', and so on.

You'd end up with something like this

64	32	16	8	4	2	1	
1	1	1	1	0	1	1	
64 +	32	+ 16	+8		+ 2	+ 1	= 123

Add the numbers together.

Convert Binary to Decimal

Using the principle mentioned earlier, lets try convert a number. Convert 01100110_2 to decimal.

First, write out your binary number and assign the place values.

Place value: 128 64 32 16 8 4 2 1
Device: ⓪①①⓪⓪①①⓪

Now wherever there is a 1, add the place values together. So you'll end up with something like this.

$$\frac{2^7}{128}\ \frac{2^6}{64}\ \frac{2^5}{32}\ \frac{2^4}{16}\ \frac{2^3}{8}\ \frac{2^2}{4}\ \frac{2^1}{2}\ \frac{2^0}{1}$$

Place value: 128 64 32 16 8 4 2 1
Device: ⓪①①⓪⓪①①⓪

$$64+32\ +\ 4+2\ =\ 102_{10}$$

So, $01100110_2 = 102_{10}$

Take a look at the video demo at

`videos.tips/data`

Convert Decimal to Binary

To convert a decimal number to binary, continually divide the number by 2. If the number divides equally then write down 0, if there is a remainder write down 1.

Lets take a look at an example.

Convert 67_{10} to binary.

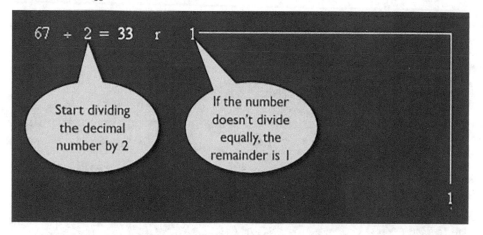

You'll end up with something like this

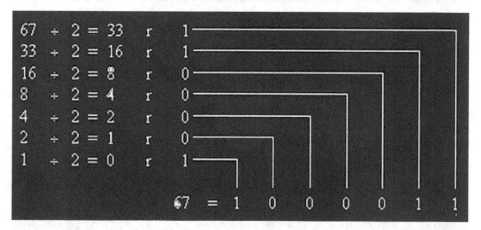

Remember, you construct the binary number reading your remainders from the last one to the first one.

Take a look at the video demo at

`videos.tips/data`

Binary Addition

Adding two binary numbers together is fairly straight forward. All you have to remember are these simple rules...

```
0 + 0 = 0
0 + 1 = 1
1 + 0 = 1
1 + 1 = 0 carry 1
1 + 1 + 1 = 1 carry 1
```

Have a look at adding these two numbers together, and apply the rules quoted above.

Working from the right to the left we get:
0+1=1, 0+1=1, 1+1=0 carry 1, 1+1+0=0 carry 1

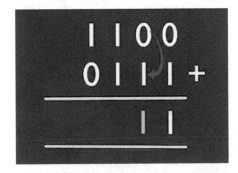

Once you work through the steps, you'll end up with something like this:

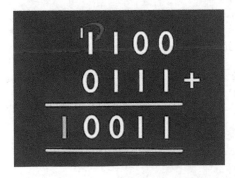

Take a look at the video demo at

videos.tips/data

Hexadecimal Numbers

Hexadecimal is used as a shorthand for binary and uses the decimal numbers 0-9 and the first 6 letters of the alphabet.

Hexadecimal	Decimal
0	0
1	1
2	2
3	3
4	4
5	5
6	6
7	7
8	8
9	9
A	10
B	11
C	12
D	13
E	14
F	15

Convert Decimal to Hexadecimal

To convert a decimal number to hexadecimal, you divide the decimal number by 16, noting the remainder.

For example, lets convert 1500_{10} to hexadecimal.

First, divide 1500 by 16

```
1500 / 16 = 93 (ignore the decimal point)
```

To find the remainder, multiply your answer by 16

```
93 x 16 = 1488
```

Subtract the answer you get from 1500

```
1500 - 1488 = 12
```

The remainder is 12

18

So, going back to our conversion we can write:

1500 / 16 = 93 r 12

Next, divide 93 by 16

93 / 16 = 5

Find the remainder

5 x 16 = 80

Subtract the answer you get from 93

93 - 80 = 13

The remainder is 13

Going back to our conversion we get

93 / 16 = 5 r 13

Write it under the previous divide step

1500 / 16 = 93 r <u>12</u>

93 / 16 = <u>5</u> r <u>13</u>

We can't divide 5 by 16 so we can stop there...

Now, read off the number starting with the answer from the last divide step, then all other remainders from the bottom up. So in this example, we get:

5, 13, 12

Remember we're converting to hexadecimal, so the number 13_{10} is D, and the number 12_{10} is C.

We end up with:

1500_{10} = $5DC_{16}$

Take a look at the video demo. Open your web browser and navigate to the following site.

videos.tips/data

Convert Hexadecimal to Decimal

To convert your hex number, reading from right to left you mark the hex number 1, 16, 256, 4096... These are called place values.

Lets convert $5DC_{16}$ to decimal.

First, write out your hex place values like this:

	16^3	16^2	16^1	16^0
Place Value	4096	256	16	1

Underneath, write down your hex number (shown in red below). Remember D_{16} is 13_{10} and C_{16} is 12_{10}. So we write the number in like this:

	16^3	16^2	16^1	16^0
Place Value	4096	256	16	1
		5	13	12

Now, multiply each number you wrote down by the place value above it.

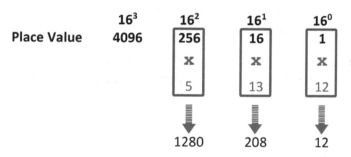

Add the results of the multiplications together.

$$\Rightarrow 1280 + 208 + 12 = 1500_{10}$$

$5DC_{16} = 1500_{10}$

Take a look at the video demo at

`videos.tips/data`

Boolean Logic

Boolean Logic or Boolean Algebra is the branch of mathematics, where values of variables are either true or false (1 and 0 respectively), as first described by 19th Century mathematician George Boole. This concept forms the basis for the development of computer electronics. Let's start with the basic logic gates.

AND Gate

An AND gate has two inputs. AND gates require both inputs to be 1 for the output to be 1. Expressed as Out = A.B

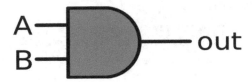

The truth table would be

A	B	Output
0	0	0
0	1	0
1	0	0
1	1	1

OR Gate

The OR gate has two inputs. OR gates require either one or both inputs to be 1 for the output to be 1. Expressed as Out = A+B

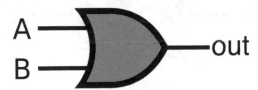

The truth table would be

A	B	Output
0	0	0
0	1	1
1	0	1
1	1	1

XOR Gate

Known as an exclusive OR gate, this gate requires the input to be either 1 or 0, not both. If both inputs are the same, the output is 0.

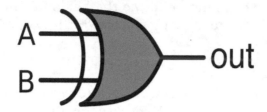

The truth table would be

A	B	Output
0	0	0
0	1	1
1	0	1
1	1	0

Expressed as Out = A⊕B

NOT Gate

A NOT gate has just one input. The NOT gate simply negates the input. So if the input is 1, the ouput is 0.

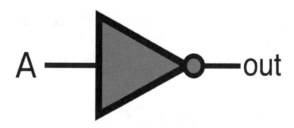

The truth table would be

A	Output
0	1
1	0

Expressed as Out = Ā

NAND Gate

Known as a negated AND gate, this gate gives the opposite output to an AND gate.

The truth table would be

A	B	Output
0	0	1
0	1	1
1	0	1
1	1	0

Expressed as Out = $\overline{A.B}$

NOR Gate

Known as a negated OR gate, this gate gives the opposite output to an OR gate.

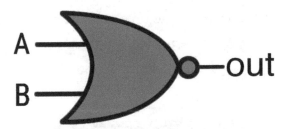

The truth table would be

A	B	Output
0	0	1
0	1	0
1	0	0
1	1	0

Expressed as Out = $\overline{A+B}$

Putting it Together

That's all well and good, but what does it all mean? Why bother? Well it turns out that these logic gates can be used to build electronic circuits. Logic gates form the building blocks used to create all sorts of electronic devices, from phones, tablets, watches, and cameras, to computers, workstations, and servers.

Using the logic gates above we can construct a little circuit that adds two numbers. Remember electronics and computers only understand 1s and 0s so we'll add two binary numbers

$$1_2 + 1_2 = 10_2$$

In the above circuit, if we set input A to 1 and input B to 1, we would get the sum of 0, carry 1.

Using the truth table, the logic on the XOR gate would be

A	B	Sum
1	1	0

On the AND gate

A	B	Carry
1	1	1

So the answer would be sum 0, carry 1. We could write that out in binary, which would be 10_2 which is 2 in decimal.

4	2	1
0	1	0

This circuit is known as a half adder and can be daisy chained together to form a full circuit.

Using these logic gates, we can start to build circuits. Say we wanted to build a simple burglar alarm system. We could have three inputs:

- A (a master on/off switch to arm and disarm the alarm)
- B (a door/window sensor)
- C (a motion sensor in a room)

Once the alarm is on, either sensor A or B can trigger the alarm.

What logic gates do we need? Well, the word 'or' in the specification above should give you a clue as to one of them: an OR Gate.

What else? The system has an on/off switch, so we can interpret this as: the system has to be ON (A) <u>AND</u> either B OR C must be ON to trigger the alarm. So we'll also need an AND gate.

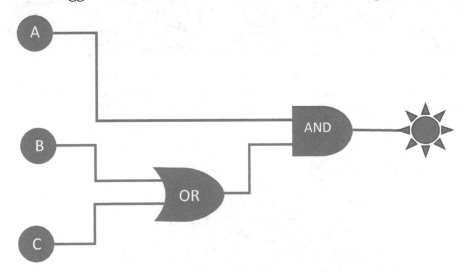

Now test the circuit with a truth table. Does it work as specified?

A (master)	B (door sensor)	C (motion sensor)	Alarm Sounds?
0	0	0	0
0	0	1	0
0	1	0	0
0	1	1	0
1	0	0	0
1	0	1	1
1	1	0	1
1	1	1	1

Harvard Architecture

In early computer systems, instructions and data were stored on punch cards or punched paper tape. The photograph below is the Harvard Mark I built by IBM in 1944.

The Mark I read its instructions from a punched paper tape. A separate tape contained data for input. This separation of data and instructions is known as the Harvard architecture.

Here in the diagram below, you can see on the Harvard architecture. There is a separate area for program instructions and another for data.

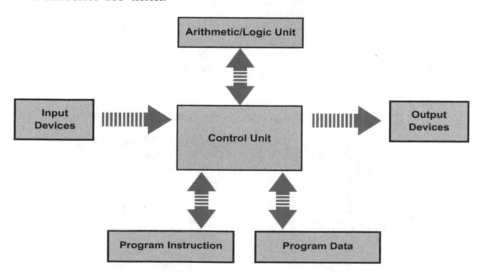

Many embedded systems in use today are based on the Harvard architecture.

Von Neumann Architecture

Von Neumann architecture is based on the stored-program computer concept, where program instructions and data are stored in the same memory.

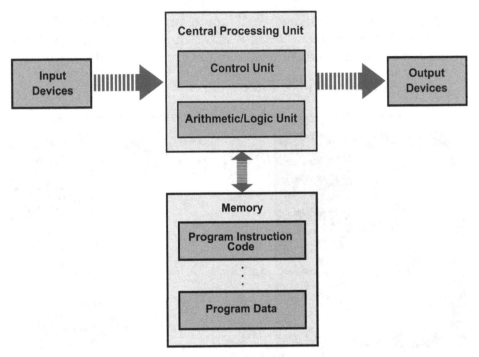

Instructions are fetched from memory one at a time and executed by the processor. Any data required by the program is fetched from memory, and any result from the execution is stored back in memory.

A processor based on Von Neumann architecture has several registers used during execution of an instruction, these are:

- Memory Address Register (MAR)
- Memory Data Register (MDR)
- Current Instruction Register (CIR)
- Program Counter (PC)
- Accumulator (ACC)

The registers and key elements of the Von Neumann architecture are used during the fetch-decode-execute cycle.

Fetch Execute Cycle

The fetch-execute cycle is the logical sequence of steps that a CPU follows to execute instructions of a program stored in memory (RAM).

To demonstrate this cycle, we'll use a very simplified CPU and memory. Many modern CPUs are far more complex, and have extra features such as caches, hyper-threading and so on.

Our little system will look something like this...

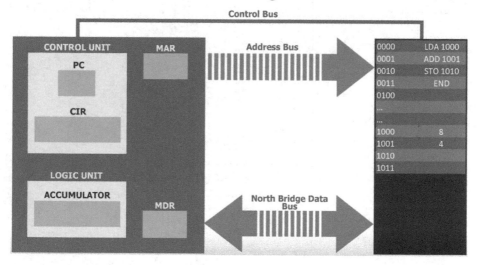

The CPU is made up of a control unit that contains the program counter (PC). This counter increments in sequence to indicate the address of the next instruction to be called from memory.

We also have a current instruction register (CIR) to hold the instruction currently being executed, and an accumulator to temporary store the result.

We have a little program that adds two numbers together and stores the result back in memory.

```
LDA 1000 # loads accumulator with data at address 1000
ADD 1001 # adds data stored at address 1001
STO 1010 # stores result in memory at address 1010
```

Lets take a look at executing the first instruction stored in memory.

```
LDA 1000
```

1. Copy PC to MAR.

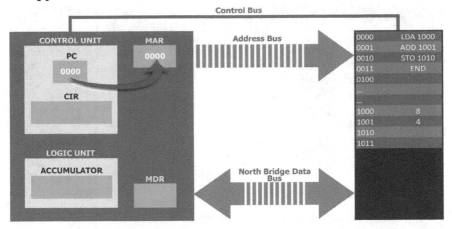

2. Copy instruction from memory (at address in MAR) to MDR.

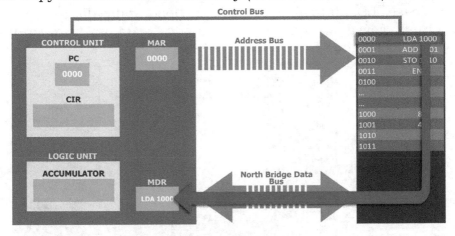

3. Copy instruction in MDR to CIR.

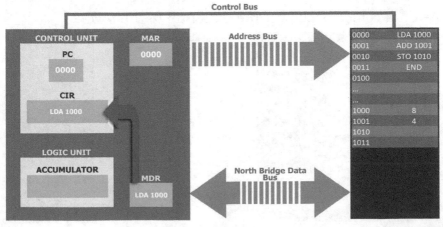

4. Increment the PC.

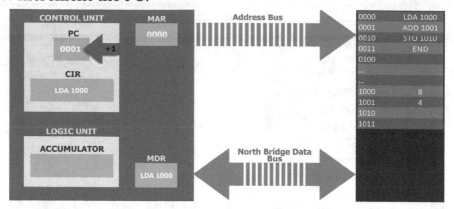

5. Decode Instruction. Strip to opcode (LDA) & operand (1000).

6. Execute instruction. What does the instruction do? Well in this case, LDA is telling the CPU to load some data from an address. How do we get data from an address? Copy the operand part of instruction from CIR to the MAR.

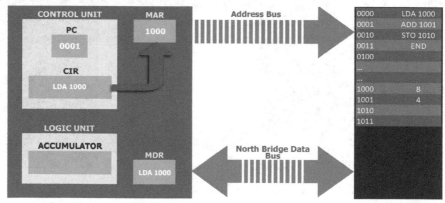

The data comes back over the data bus to the MDR.

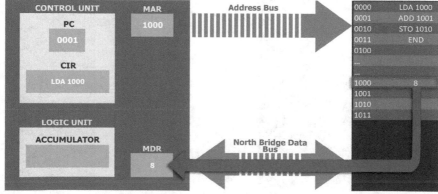

Now we have the data, we can carry out the instruction. Add 8 to the accumulator.

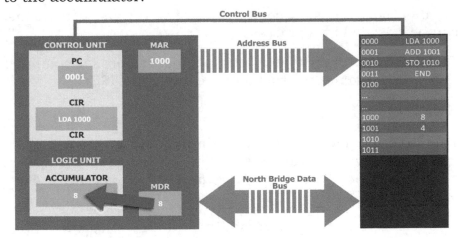

Got it? Try running the next instruction. `ADD 1001`

Here's the sequence of steps to help you.

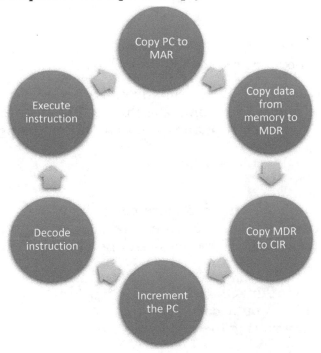

If you get stuck, take a look at the video demos to see the sequence in action. Open your web browser and navigate to the following website.

`videos.tips/architecture`

Instruction Sets

An instruction set is the complete set of all the instructions that can be executed by a processor.

There are two main types: reduced instruction set computer (RISC) and complex instruction set computer (CISC).

RISC

The RISC architecture uses a simpler set of more general instructions where each instruction is usually executed in one clock cycle. *Note the term 'reduced' does not mean fewer instructions.*

Lets look at an example. To add two numbers together, you'd use general instructions such as LDA for load, ADD for addition and STO for storing the result back to memory.

```
LDA 1000
ADD 1001
STO 1010
```

These are known as reduced instructions.

RISC processors are used in many game consoles such as the game cube, playstation, and xbox using IBM's PowerPC chip, as well as many smart phones such as Android using the ARM processor.

CISC

The CISC architecture uses more complex instructions, where a single instruction can execute several operations such as a load data from memory, perform an operation, and store result in memory.

So going back to our ADD instruction, using the CISC architecture the instruction would carry out the loading of the data, the operation, and storing the result back to memory.

```
ADD 1000, 1001
```

This is known as a complex instruction.

Most modern computers such as PCs, Macs, Laptops, and tablets use the CISC architecture on Intel and AMD processors.

Data Compression

Data compression involves encoding data using fewer bits than the original, thereby reducing the overall size. Data is usually compressed for storage or transmission over a network.

A method of data compression can be either lossy or lossless.

Lossy & Lossless

Lossy compression methods remove unnecessary or redundant data and is used to compress multimedia files such as photos, videos and audio. Some examples: JPEG images, MP4 videos, MP3 audio

MP3s are compressed using a technique called psycho acoustic compression which is based on how humans hear sounds. For example, a quiet sound immediately following a loud sound is often not heard, so the quiet sound can be removed. Also any audio that falls outside the human audible range of 20Hz to 20Khz can be removed.

Lossless compression methods reduce the file size without removing any data, so the file can be fully reconstructed when decompressed. Some examples are: ZIP

Huffman Coding

Huffman coding is a lossless data compression algorithm based on the frequency of occurrence of a data item. Codes of different lengths are assigned to characters based on the frequency of occurrence. Smaller codes are assigned to characters that have the highest occurrence.

Lets look at an example. Let's encode the word LEMMONS

First, create a table showing the frequency. Order the table by frequency from lowest to highest.

Char	Freq
S	1
N	1
O	1
E	1
L	1
M	2

Now construct a Huffman tree. Create leaves for each of your characters using the table above.

Take the first two characters from your list (S & N). Pair them together, add the frequencies. 'S' appears once, 'N' appears once, so frequency is 2. Put them back in the list. *Remember, the frequencies must be in order, the combined frequency is 2, so it goes after 'M' whose frequency is 2.*

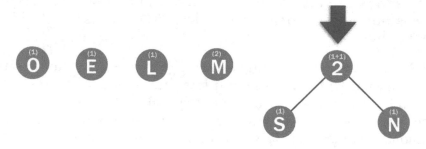

Take the next two characters from your list (O & E), pair them, add the frequencies, put them back in the list in order.

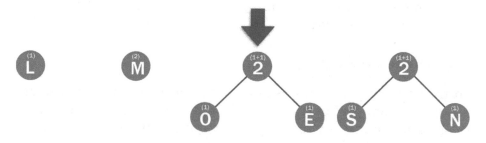

Take the next two characters from your list (L & M), pair them, add the frequencies. Add the node back into the list in order. The frequency is 3, so it goes at the end.

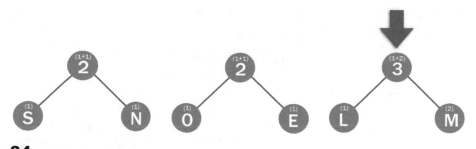

Now, take the next two from your list - nodes (S&N and O&E). Pair them, add the frequencies (2 + 2 = 4).

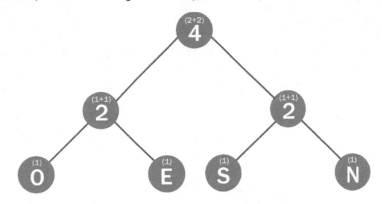

Put back in the list. Frequency is 4 so it goes at the end after node (L&M).

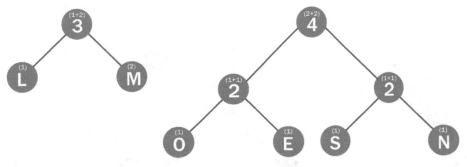

Take the next two from the list, pair them, add the frequencies (3 + 4 = 7).

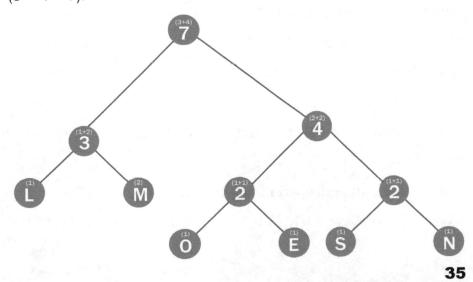

Now to encode the data you need to label the branches.

- Left branches are 0
- Right branches are 1

Each character that appears in the tree is assigned a unique code (a sequence of 0s and 1s) obtained by following the path from the root of the tree to the leaf containing the character.

So to encode a character you traverse the tree from the top node. Note the pattern of 1s & 0s until you reach the character. For example, E would be encoded as 101

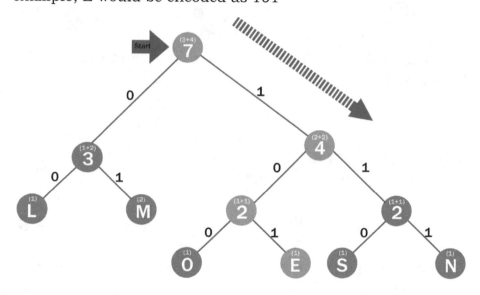

L would be encoded as 00

M would be encoded as 01

O would be encoded as 100

N would be encoded as 111

Using Huffman, LEMMONS would be encoded as

00 101 01 01 100 111 110

The original data encoded in ASCII would be:

01001100 01000101 01001101 01001101 01001111 01001110 01010011

So instead of needing 56 bits (7 bytes), with Huffman compression you only need 18bits. You'd also have to store the tree as a translation table in order to decompress the data.

Decoding the tree, you work the opposite way.

Decode 111. To do this we go to the tree. Start from the top and traverse the tree. Turn left for a 0, turn right for a 1.

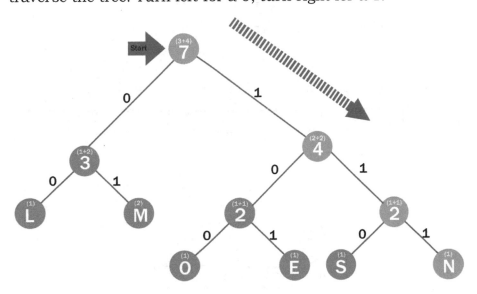

111 is N. What about 101?

Take a look at the 'huffman' coding section of the video demos. Open your web browser and navigate to the following website.

`videos.tips/data`

Run Length Encoding

Run length encoding is a lossless data compression algorithm usually used to compress repetitive data. The aim is to reduce the number of bits used to represent a set of data.

The compression process involves counting the number of consecutive occurrences of each character (called a run).

Let's have a look at an example.

aaaaaaaabbbbbbcc

We can encode this as

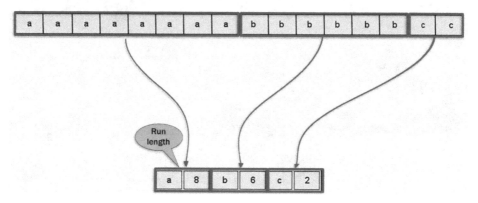

aaaaaaaabbbbbbcc would be encoded as a8 b6 c2

Encoded in ASCII the data stream would be 16 bytes.

```
01100001 01100001 01100001 01100001
01100001 01100001 01100001 01100001
01100010 01100010 01100010 01100010
01100010 01100010 01100011 01100011
```

Using RLE, the data stream would be 6 bytes.

```
01100001 00001000 01100010 00000110
01100011 00000010
```

RLE is particularly well suited to palette-based bitmap images such as computer icons.

Data Encryption

Data encryption is a method of keeping data secure in storage or when transmitted, where the data is encoded using an encryption key, and can only be decrypted by a user with the correct key.

The un-encrypted data is called plaintext, and the encrypted data is called a ciphertext.

To demonstrate how encryption works, we'll use a very simple technique called a Caesar cipher.

With Caesar ciphers, each letter in the plaintext is replaced by a letter a fixed number of positions down the alphabet.

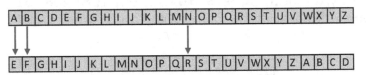

So to encrypt the word BANANA, you would get FERERE

B A N A N A

F E R E R E

Another way to encrypt data is to use a Vernam cipher. With this scheme, each plaintext character is mixed with one character from a key. This is achieved by applying the logical XOR operation to the individual bits of plaintext and the random key. See page 22 for the XOR truth table.

So to encrypt the word BANANA

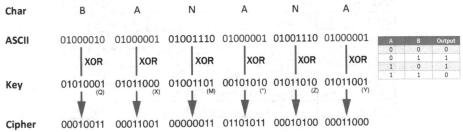

Plaintext would be (BANANA):
01000010 01000001 01001110 01000001 01001110 01000001

Cipher would be (‼├└k¶↑):
00010011 00011001 00000011 01101011 00010100 00011000

Data Storage Units

Much like length can be measured in metres and kilometres or weight in grams and kilograms.

For example, you'd have 1000 meters in a kilometre, or 1000 grams in a kilogram.

Computer storage is measured in the same fashion using bytes and kilobytes.

So, for example, you'd have 1000 bytes in a kilobyte.

Technically 1 kilobyte is exactly 1024 bytes because computers use binary to count (2, 4, 8, 16, 32, 64, 128, 256, 512, 1024), not the decimal system (1, 10, 100, 1000, 10000, 100000) like we do.

Bit	Either 1 or 0
Byte	8 bits
KiloByte	1024 Bytes
MegaByte	1024 KiloBytes
GigaByte	1024 MegaBytes
TeraByte	1024 GigaBytes

Here's a summary of the different units you'll most likely come across when using computers.

Unit	Size in Bytes	Abbr
KiloByte	1,024	KB
MegaByte	1,048,576	MB
GigaByte	1,073,741,824	GB
TeraByte	1,099,511,627,776	TB
PetaByte	1,125,899,906,842,620	PB
ExaByte	1,152,921,504,606,850,000	EB
ZettaByte	1,180,591,620,717,410,000,000	ZB
YottaByte	1,208,925,819,614,630,000,000,000	YB

To give you an idea, music and photos can be a couple of mega bytes each but can take up a few hundred gigabytes if you have a lot.

Large documents can be around 40 megabytes, or about 50 kilo bytes if they are short.

Different types of storage media have different sizes. Here are a few examples of data storage sizes.

Floppy disc	*1.44MB*
CD-ROM	*800MB*
DVD	*4.7GB*
USB memory stick	*16GB*
Hard drive	*1TB*

Floppy discs, which are somewhat obsolete these days, held 1.44MB of data, probably the size of a small word document.

USB memory sticks vary in size and can be anywhere between 1GB and 256GB.

A hard disc drive can store over 1 TB of data.

Chapter 2

Hardware Components

Computer systems are made up of various different hardware components such as a central processor (CPU), memory (RAM), storage space (HDD) and so on. This is called internal hardware and usually plugs into a main board called a motherboard.

Devices that sit outside are called peripherals and include printers, scanners, keyboards, mice, cameras and so on.

There is also removable storage such as memory cards, USB sticks and external hard drives that are designed to be portable.

Let's start by taking a look at internal hardware.

Types of Hardware

On a desktop computer, the case houses all the internal hardware, such as CPU, RAM and Hard discs. Peripherals sit outside the case, and the Operating System runs on the internal hardware.

Internal Hardware

These are the three primary components. We'll go into more detail in later chapters, but for now here is an overview.

These are all mounted onto a main circuit board, called a motherboard.

Basic Peripherals

The most common peripherals are

- Monitors

- Printers

- Keyboards

- Mice

- Scanners

- Cameras

These all sit outside the case and connect to your computer via USB cable or similar.

Some familiar ones are pictured below. Here, you have a computer monitor and printer. These are sometimes called output devices.

A keyboard and mouse used on standard PCs, laptops, servers and so on. These are sometimes called input devices.

Primary Storage Devices

Primary storage, also known as main memory, or internal memory, is memory that is accessed directly by the CPU.

Random Access Memory (RAM)

Computer memory is made up of silicon chips and is the computer's working area. This is where software instructions and data are stored.

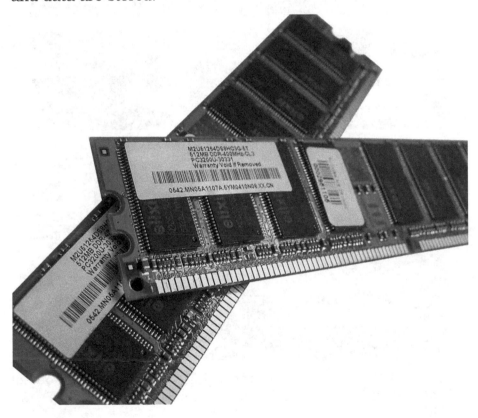

For example if you are typing a document in Microsoft Word, both Microsoft Word and your document are loaded into and stored in the computer's memory while you are working on it. This is not to be confused with the Hard Disc.

DRAM or Dynamic RAM needs to be constantly refreshed. SRAM or static RAM is a lot faster because it doesn't need to be refreshed.

Standard system RAM is DRAM or sDRAM. SRAM is reserved for cache memory and is mounted onto the motherboard.

To confuse things even more, RAM comes in different forms: DDR, DDR2, DDR3 and DDR4.

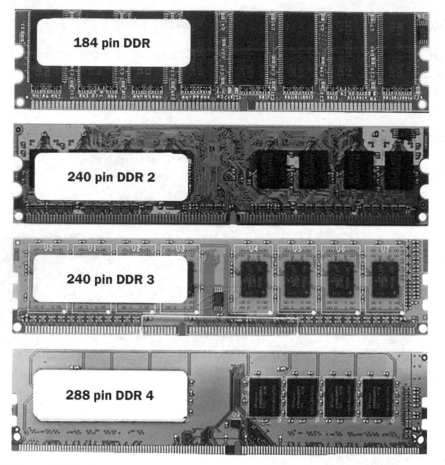

DDR and DDR2 are old now and being phased out in favour of DDR3 and DDR4. Most of the latest PCs will use DDR3 & DDR4 RAM.

Notice the positions of the cut out notches along the bottom of the DIMM indicated in red on the photograph above. This is to make sure only the correct RAM fits in the slot on the motherboard.

RAM speed is also measured in MegaHertz (MHz) and you'll likely see this when looking at buying RAM.

Laptops have their own type of memory. It's more or less the same except for the physical size. These memory modules are called SO-DIMMs

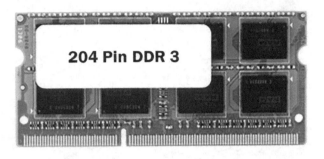

204 Pin DDR 3

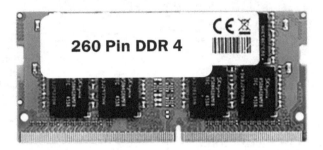

260 Pin DDR 4

When buying memory, you need to find out what memory modules your computer's motherboard takes. You should be able to find this in the documentation that came with your computer.

So looking at the spec of my computer, shown below, I need a DDR4 memory module that runs at about 2133MHz.

Slots	Four DIMM sockets
Type	DDR4
Speed	• 2133 MHz • Up to 2400 MHz with XMP (optional)
Configurations supported	4 GB, 8 GB, 16 GB, 32 GB, and 64 GB

Also reading further down, I can get 4, 8,16,32 or 64GB modules. For this example I am going to add 16GB

So when searching, I'll look for something like the one shown below, and make sure the module is DDR4 and runs at least 2133MHz as it said in the PCs spec above.

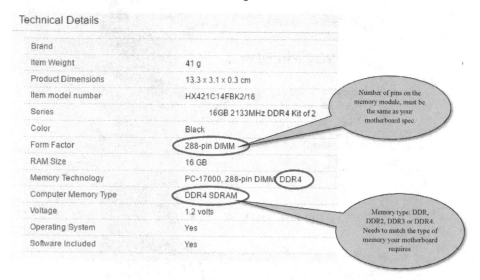

Technical Details	
Brand	
Item Weight	41 g
Product Dimensions	13.3 x 3.1 x 0.3 cm
Item model number	HX421C14FBK2/16
Series	16GB 2133MHz DDR4 Kit of 2
Color	Black
Form Factor	288-pin DIMM
RAM Size	16 GB
Memory Technology	PC-17000, 288-pin DIMM DDR4
Computer Memory Type	DDR4 SDRAM
Voltage	1.2 volts
Operating System	Yes
Software Included	Yes

Number of pins on the memory module, must be the same as your motherboard spec

Memory type: DDR, DDR2, DDR3 or DDR4. Needs to match the type of memory your motherboard requires

You will also need to open the case and find out how many memory slots are available.

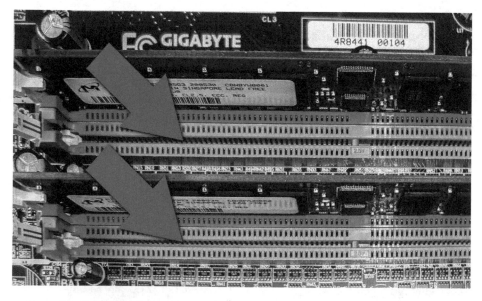

This one has two spare and two in use. Slide the DIMM into a vacant slot, it will only go one way, then gently push down until it clicks into place.

Read Only Memory (ROM)

Read Only Memory is non-volatile memory or storage containing data that cannot be changed.

Read Only Memory is useful for storing a program that very rarely change. An example is the BIOS program needed to start a PC, sometimes known as firmware.

Cache Memory

The cache is usually an extremely fast memory chip that stores data so that subsequent requests for that data can be served faster.

Data read from a hard disc drive can be stored in a cache, so when a program requests that data again, it can be read from the cache instead of from the hard disc drive.

Volatile and Non-Volatile Memory

Volatile memory requires a power source to maintain the stored data. Once the power is cut, the data is lost. The computer's main memory, the RAM, is known as volatile memory.

Non-volatile memory can retain the stored data even when the power is turned off. The computer's hard disc drive is an example of a non-volatile memory store. Non-volatile memory is used for permanent storage and backup.

Secondary Storage

Secondary storage, also known as auxiliary storage is memory that is used to permanently store computer software and data. When you install an Operating System, an application, or save a file, these are stored on a secondary storage device such as a hard disc drive. Secondary storage devices are divided into three types: magnetic, solid state, and optical.

Magnetic storage devices such as hard disc drives use a magnetic field to magnetise sections of the disc to store data. These devices tend to be large in capacity and cheap.

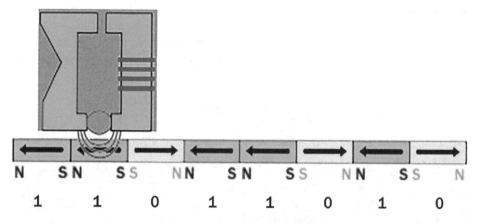

Solid state devices use flash memory to store data. These devices tend to be small in capacity, expensive, and are fast.

Optical devices use a laser to read data from a CD, DVD or Bluray disc. You can also write data to a disc, however this is usually permanent and can't be erased.

Hard Disc Drive (HDD)

The hard disc (also called a hard drive) is like a filing cabinet and permanently stores all your documents, photographs, music, your operating system (such as Microsoft windows) and your installed software (such as Microsoft word).

When you start up an application such as Microsoft word, the Microsoft word software is loaded up off the hard disc into the computer's primary memory (or RAM), where you can work on your documents.

The capacity or size of the disc is usually measured in Gigabytes or Terabytes, eg 500GB, 1TB, 4TB.

The disc spins at 7200rpm on most modern hard disc drives and connects to your computer's motherboard using a SATA cable. Some cheaper drives run at 5400rpm and can be slow on modern computers, so keep this in mind when buying.

These are called internal drives, and must first be formatted by the Operating System before it can store data.

PCs use 3.5" drives, while laptops use 2.5" drives.

Inside the hard drive, you'll see a stack of double sided disc platters with an actuator arm containing a electromagnetic read/write head that hovers microns above the surface of the disc.

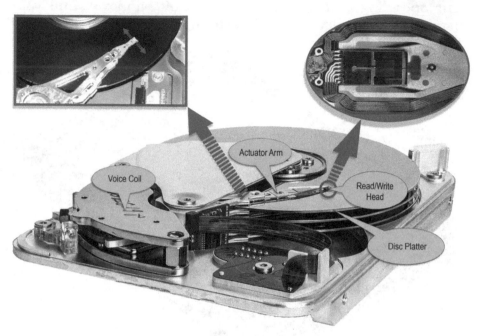

As the disc platters rotate, the actuator arm moves back and forth positioning the head on the correct track/sector of the disc in order to read or write data.

Solid State Drive (SSD)

These drives perform much like a traditional Hard Disc but are extremely fast and also expensive. SSDs have no moving parts and are composed of non-volatile NAND flash memory, which is a type of memory that retains data even if the power is turned off.

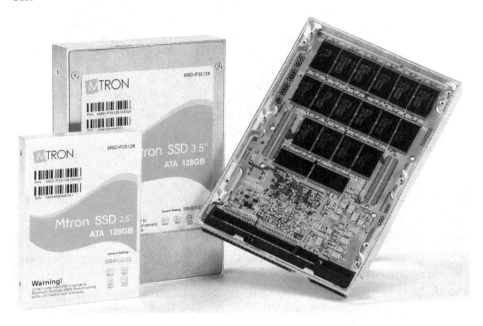

These drives are being used in smaller laptops/notebook computers and on some tablet computers where you don't require large amounts of storage space.

These drives can also be useful where you need fast data transfer rates, such as for video editing, playing games and recording audio/video.

SSD drives can be over 2TB, however large capacities are expensive. The average size in use is about 250GB.

SSDs are slowly becoming more popular and some of the more modern computers come with a SSD installed.

These drives give a tremendous speed boost to the computer, allowing applications to start up in seconds, as well as allowing the operating system to start much quicker than with a traditional hard disc drive (HDD).

CD/DVD/Blu-ray Drive

Another dying breed, this drive allows you to play CDs, watch
DVDs or Blu-Ray movies that come on a disc.

The surface of the disc is marked with pits. This is how the data
is encoded onto the disc. Each time a pit is encountered, the
laser beam is not reflected and is interpreted as a 0. If there is
no pit, the laser beam is reflected and is interpreted as a 1.

Here we've removed the metal casing so you can see the components inside.

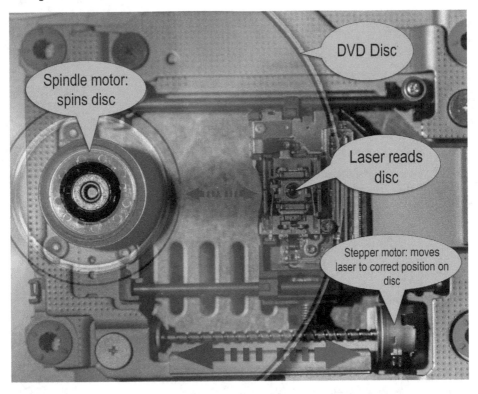

A stepper motor moves the laser head assembly back and forth to the correct track on the disc, while the spindle motor rotates the disc at a constant speed.

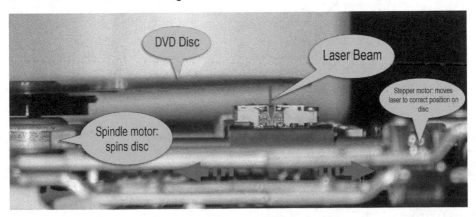

Here you can see a low powered laser beam focused onto the surface of the disc.

Lets take a closer look inside the laser head assembly. Here, we've broken the device down into its components.

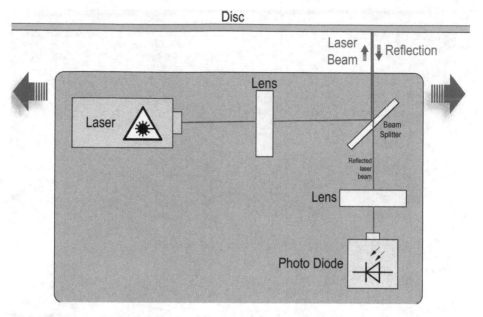

The green line across the top of the diagram represents the disc. This spins around when the drive is in use.

The grey box contains the laser and photo diodes used to read the pits in the surface of the disc.

The laser fires a beam of light that is focused through a lens onto the surface of the disc.

If the laser beam hits a pit, no light is reflected back to the photo diode and is encoded as a 0.

If the laser beam hits the surface of the disc where there is no pit (called a land), the light is reflected back to the photo diode, and is encoded as a 1.

Now using the binary code, data can be encoded in these pits on the surface of the disc.

Have a look at the video demos to see how these devices work internally. Open your web browser and navigate to the following website

www.elluminetpress.com/computer-systems

Memory Cards

Many laptops and tablets now have memory card readers built in. The most common memory card is the SD Card. This can be a full sized SD card or a Micro SD card.

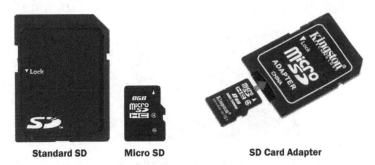

Standard SD   Micro SD   SD Card Adapter

Standard SD cards are commonly used in digital cameras, and many laptops have standard size SD card readers built in. Tablets, phones, and small cameras usually use micro SD cards.

You can get an SD Card adapter if your SD card reader does not read Micro SD cards.

There are various types of SD cards available, each are marked with a speed classification symbol indicating the data transfer speed. If you are merely storing files, the data transfer speed doesn't really matter as much, however, if you are using the card in a dash cam or digital camera, the faster data transfer speeds are necessary. To be safe, the higher the speed the better. You can see a summary in the table below.

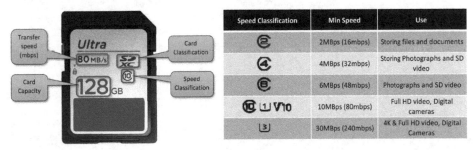

Speed Classification	Min Speed	Use
②	2MBps (16mbps)	Storing files and documents
④	4MBps (32mbps)	Storing Photographs and SD video
⑥	6MBps (48mbps)	Photographs and SD video
⑩ U1 V10	10MBps (80mbps)	Full HD video, Digital cameras
3	30MBps (240mbps)	4K & Full HD video, Digital Cameras

SDHC stands for "Secure Digital High Capacity", and supports capacities up to 32 GB.

SDXC stands for "Secure Digital eXtended Capacity", and supports capacities up to 2 TB.

Chapter 2: Hardware Components

SDUC stands for "Secure Digital Ultra Capacity", and supports capacities up to 128 TB.

These cards are usually read with a card reader. Most tablets and smart phones have these built in, however you can buy USB card readers that plug into your computer like the one below.

Most modern laptops or tablets have a built in reader. It is usually on either of the side panels or the front panel.

The card will show up as another drive in file explorer in Windows. Click on the drive to see the contents of the card.

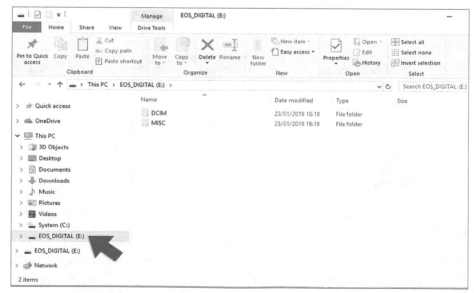

USB Flash Drives

Also called memory sticks, these little things plug into the USB port on your computer or laptop and allow you to copy files onto them, much like a hard disc.

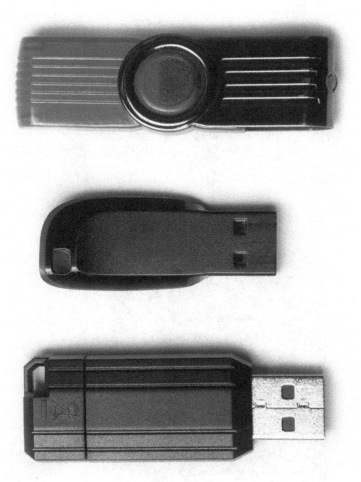

These USB flash drives are made up of flash memory and are formatted using FAT32, or exFAT file systems. This allows them to be read by a wide range on Operating Systems.

They can be used for backup and transport of data such as photos, documents etc. You can also use them to boot from, allowing you to install operating systems on a computer.

These drives come in a variety of shapes and capacities ranging from 2GB up to 512GB

External Hard Disc

External hard discs are hard disc drives that sit outside the computer and plug into your USB port - usually USB-2 and USB-3 and USB-C.

These drives are designed for portability and are usually small enough to fit in your pocket.

They are great for backup and transport of data such as photos, documents, videos etc.

External hard disc drives can range up to about 4TB and beyond.

NAS Drives

NAS drives, sometimes called Network Attached Storage, allow you to store and backup files to a central point on a network.

All the machines on the network can access files on the NAS drive. Shared data can either be set to private for one particular machine or shared publicly for all machines to see.

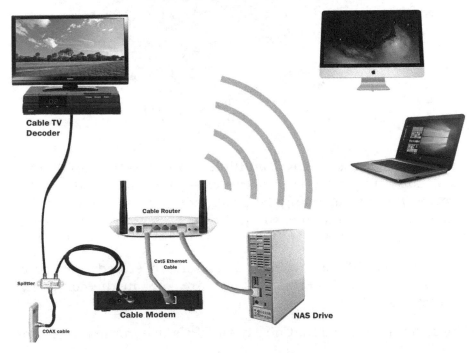

These devices make great backup strategies and come with software you can install on your computer to automate backups at certain times. You will need a few terabytes of storage on your NAS.

CPU / Processor

The CPU or processor is the brain of the computer and responds to all the commands you give the computer. It is one of the primary factors in determining the power of the system. Measured in Gigahertz, the higher the number, the more powerful the processor.

Modern processors have multiple cores. You might see a dual core or quad core processor. A core is an independent processing unit, meaning the processor can execute more than one instruction at a time, so the more cores your processor has, theoretically the faster it is.

The job of the CPU, is to execute a sequence of stored instructions called a program. The instructions are kept in the computer's memory (or RAM).

There are four steps that nearly all CPUs use in their operation:

fetch, decode, execute, writeback

Processors running at 2Ghz or more, do this billions of times a second.

Types of CPU

There are two main CPU manufacturers: Intel and AMD. There are various different CPUs, each running at different speeds and manufactured for different computers such as laptops, desktops, or tablets. Each manufacturer has their own specs, names and series for their CPUs.

This is where things get a bit confusing. There are a lot of different processors out there with different numbers and series. However it all boils down to a few numbers to take note of, I've highlighted these in bold. Lets take a look at some common ones.

Intel Pentium / Celeron. These chips are common in cheap laptops, and offer the slowest performance, but can handle tasks such as web browsing, email and document editing. You'd be better off spending a bit extra and going with a Core i3 or i5.

Intel Core i3. Performance is about entry level for basic computer usage such as web browsing, email, social media, word processing, music and looking at a few photos.

AMD A, FX or E Series. Found on low-cost laptops, AMD's processors provide decent performance for the money that's good enough for web browsing, internet, email, streaming films or tv, photos and music, as well as word processing etc. For example:

> AMD **A6**-9220 APU 2.5GHz
> AMD Quad-Core Processor **FX**-9830P

Intel Core i5. If you're looking for a mainstream laptop with the best combination of price and performance, get one with an Intel Core i5 CPU. Always make sure the model number ends with a 'U', HK, or 'HQ' - these offer better performance. For example:

> Intel Core i5-7200**U**
> Intel Core i5-7300**HQ**

Also the higher the number after 'i5', eg 7300, the better the performance.

Intel Core i7. The successor to the Core i5.

AMD Ryzen Series. High powered chips from AMD designed to compete with Intel Core i5 and Core i7. Great alternative to Intel chips and good for gaming and high powered laptops. For example: AMD 8-Core **Ryzen R7** 1700.

Inside the CPU

This is a die shot of an old Pentium CPU. Note the marked sections.

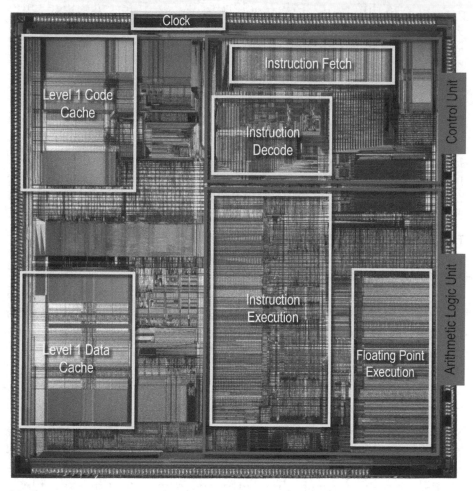

At the top you can also see the clock. The clock sets the rate at which the computer executes instructions and is usually measured in GigaHertz (GHz). This is how CPUs are rated.

On each clock cycle, the control unit fetches and decodes an instruction; this is called the instruction cycle.

Then the arithmetic logic unit executes the instruction and stores the data; this is called the execution cycle.

This cycle happens billions of times a second. So a 2 GHz CPU can execute 2 billion instructions a second.

Other Internal Components

Computers also contain other components that perform different tasks. Two common ones are a sound card and a video card. Other internal components include ethernet and WiFi cards for network connectivity.

Sound Card

A sound card, also known as an audio card, is an internal expansion card that facilitates the input and output of audio signals to and from a computer. This allows multimedia applications such as music, video, audio, presentations, and games to play sound through a speaker or sound system.

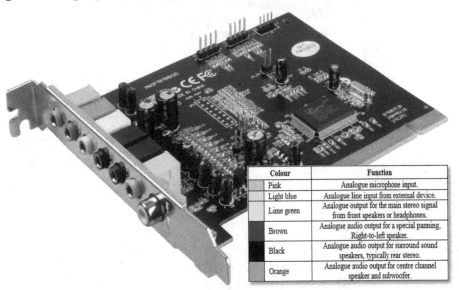

Colour	Function
Pink	Analogue microphone input.
Light blue	Analogue line input from external device.
Lime green	Analogue output for the main stereo signal from front speakers or headphones.
Brown	Analogue audio output for a special panning, Right-to-left speaker.
Black	Analogue audio output for surround sound speakers, typically rear stereo.
Orange	Analogue audio output for centre channel speaker and subwoofer.

Sound cards are usually integrated into most modern motherboards, using basically the same components as a plug-in card.

The best plug-in cards, which use better and more expensive components, can achieve higher quality than integrated sound and are usually used in higher end applications such as audio production, music composition and video editing.

Some sound cards have more specialist connections such as digital output for connecting to sound systems and amplifiers.

Video Card

The video card or graphics card is responsible for processing video, graphic and visual effects you see on your monitor. The graphics card is also known as a GPU (graphics processing unit).

Most video cards offer various functions such as accelerated rendering of 3D scenes and 2D graphics, MPEG-2/MPEG-4 decoding, TV output, or the ability to connect multiple monitors.

Most modern motherboards have video cards integrated into them, eliminating the need for a plug-in card. However, integrated video cards are not usually as high quality as plug in cards. This makes plug in cards more suited to high end video production, graphics processing and video games.

Some plug in video cards have more specialist connections such as HDMI, DVI, S-Video or Composite for connecting to high end televisions, projectors and monitors.

An example of a graphics card is the GeForce GTX 1080 Ti.

Expansion Slots

Video cards and sound cards plug into expansion slots on your motherboard. On most modern motherboards, the expansion slots are called PCI express (PCIe).

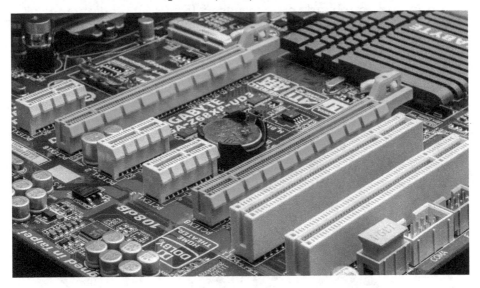

The 'PCIe x 1' slots are for smaller devices such as ethernet and wifi network controllers or modems etc.

The 'PCIe x 16' slots are extremely fast and are for high end graphics cards like the one in the previous section.

The PCI slots are for sound cards and other types of cards.

The Motherboard

All the components connect to a large circuit board called a motherboard and is the main circuit board found in desktop and laptop computer systems. It holds many of the crucial components, such as the processor (or CPU) and memory (or RAM), and provides connectors for other peripherals.

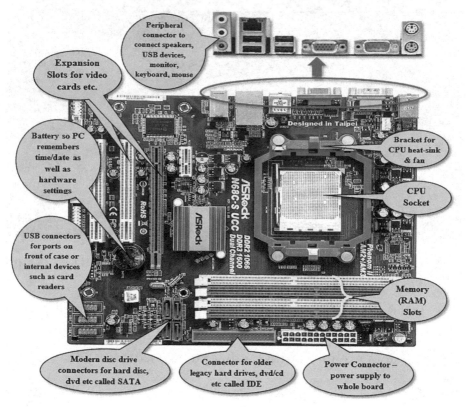

A typical desktop computer has its processor (CPU), main memory (RAM), and other essential components connected to the motherboard. Other components such as storage (hard disc, dvd drive) can be connected to the drive connectors on the motherboard using cables, as can be seen in the photograph opposite.

Cards for video display and sound may be attached to the motherboard, and plug into the expansion slots. In modern computers it is increasingly common to integrate some of these devices into the motherboard itself namely video and sound cards.

The Chipset

A chipset is the circuit that orchestrates the flow of data to and from key components of the motherboard, such as CPU to memory, as well as the data flow to and from hard drives, external drives, and peripherals.

North & South Bridge

The north bridge connects the CPU to the RAM and the PCI Express Lane.

The south bridge connects the CPU to the slower devices such as USB ports, hard drives, external drives, printers wifi/network cards, and other peripherals.

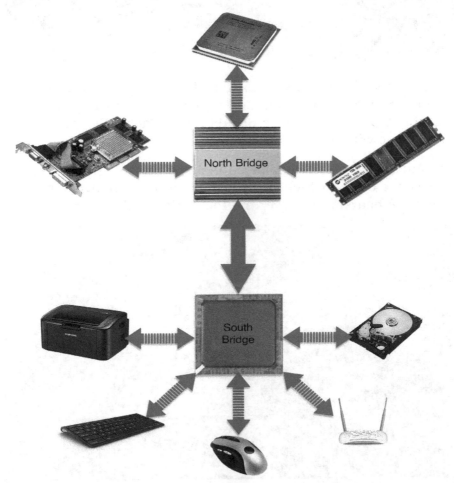

CPU Socket

This is a master socket mounted on the motherboard for housing the CPU. These sockets are known as Zero Insertion Force (or ZIF) meaning the CPU drops into the socket and secured with a locking lever. This allows the CPU to be replaced or upgraded.

There are various types of socket. For example, the Intel Core i5 uses Socket LGA 1156, Core 2 Duo uses socket LGA 775, and the AMD Ryzen 5 uses Socket AM4. Each CPU type is only compatible with its own socket type.

The microchip itself is packaged inside a durable plastic, metal, or ceramic casing which protects the chip and allows for better cooling. A heatsink fits over the top of the CPU to remove heat.

The BIOS

The BIOS is a small program stored in permanent ROM and is soldered directly onto the motherboard.

BIOS Stands for Basic Input Output System and is responsible for checking hardware components at bootup, called a power on self test (POST).

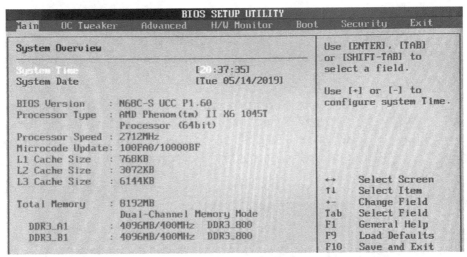

The BIOS also sets the boot sequence and will list drives the system will search in order, for an Operating System such as Windows.

The settings configured using the BIOS are stored in a small memory chip called CMOS which is powered by a small button battery. This allows the system to store the configuration along with the current time and date even when the power to the computer is turned off.

UEFI is expected to eventually replace BIOS.

Computer
Peripherals

Computer peripherals are essentially anything that connects to the computer.

This can be input devices such as keyboards, mice and scanners; output devices such as monitors and printers; or storage devices such as hard discs, DVD and flash drives.

All these components connect to the computer using a variety of different connectors and cables; whether it's USB to connect a printer or HDMI to connect a computer screen or projector.

Let's start by taking a look at the most common of all peripherals, the printer.

Printers

Printers come in two main types: inkjet and laser.

Inkjet Printer

These printers are good for the average user who just wants to print some letters or other documents and the odd few photographs. They are generally slower printers and are not suitable for printing documents with a large number of pages.

These printers can also print on labels, envelopes and specialist presentation paper (good for greetings cards if you want to print your own).

The only issue I find with inkjet printers is the ink tends to dry up if you don't print out regularly. So make sure you print out something at least once a week to keep the ink from drying up.

Inkjet printers work by forcing tiny droplets of ink in a pattern to form an image.

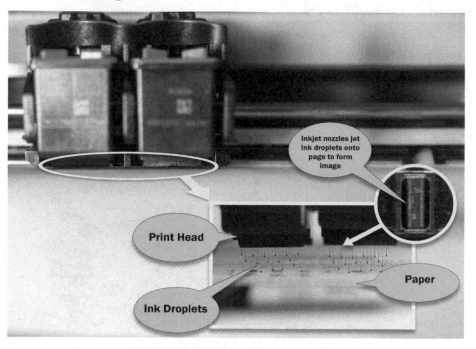

Laser Printer

Laser printers produce high quality prints very quickly and are suited to high volumes of printouts. These printers are good if you do a lot of printing, for example, if you run a business or have a family that all want to print out from their own computers/laptops at the same time.

These printers print can in black and white or colour and work by burning ink (called toner) onto the page. The toner cartridges are expensive to buy but last a lot longer than ink jet cartridges.

Laser printers use a laser beam to create the image to be printed on an electro-statically charged rotating drum.

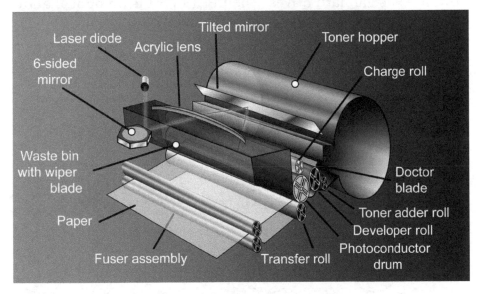

Toner particles are then picked up by the drum, creating a very sharp image.

The drum then transfers this image onto the paper by direct contact as the paper passes over it.

Finally this image is fused to the paper by the fuser unit, which heats the paper melting the toner and pressing it onto the page.

A colour laser printer will go through this process for each of the three primary colours (yellow, cyan, magenta) and black to create a colour image on the page.

Other Peripherals

Other peripherals include scanners, cameras, game control pads, virtual reality headsets, and so on.

Most external peripherals connect to the computer through a data port using a USB cable. Some older peripherals connect using firewire, but most use USB. eSATA ports can connect certain external hard drives. The SPDIF port is an optical audio port that allows you to connect high end sound systems.

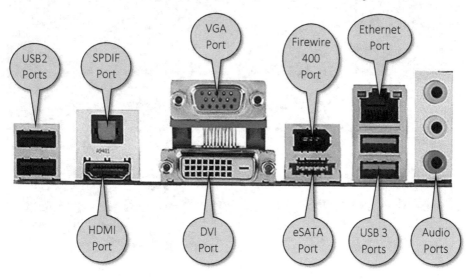

Monitors, projectors and other screens are connected using either HDMI, VGA or DVI.

3.5mm phono jacks are used to connect microphones and speaker systems to a computer. The blue port is called 'line in' and is used to connect external audio devices for recording. The green port is called 'line out' and is used to connect speaker systems and headphones. The pink port is called 'mic in' and is used to connect microphones.

Data Ports

Data ports allow you to connect devices to your computer. The most common data ports are USB, Ethernet.

USB

USB stands for Universal Serial Bus and is a universal connection used to connect all different types of peripherals to your computer as easy as possible using the same connection type.

USB 3.0, shown below left, was released on 12 November 2008, with a data rate of around 4 Gbps and is much faster than USB 2.0.

USB 2.0, shown below right, was released in April 2000, with a maximum data rate of 480 Mbps.

USB 3.0 ports are colour coded in blue, while USB 2.0 ports are colour coded in black.

The smaller USB pictured below left is called micro USB and the one next to it is called mini USB.

USB-C

USB C is a newer USB standard with a data rate between 10 & 20 Gbps and is much faster than USB 3.0.

The USB C port is different than the previous two and looks like this:

The USB C plug is a double sided connector meaning it can be plugged into a USB C port either way up.

Many new tablets and phones are now starting to include the USB-C port.

Some of the newer laptops are also including USB-C ports.

Ethernet

Also known as RJ45, Ethernet is used to connect a computer or laptop to a network and to the internet.

FireWire

Also known as IEEE 1394 or iLink, this port was widely used in digital camcorders and most of them that recorded onto tape included a firewire interface .

There were two versions; Firewire 400 (on the left) and Firewire 800 (on the right).

FireWire 400 transfers data at about 400Mbps, Firewire 800 transfers data at about 800Mbps

ThunderBolt

Thunderbolt ports are used for peripherals that require extremely fast data transfers and have been known to support speeds of up to 10Gbps.

This port is also used on Apple Mac computers as a mini display port for connecting to monitors and projectors.

eSATA

eSATA cables connect to some types of high speed external portable hard drives. The eSATA cable cannot transmit power, unless you use eSATAp (powered eSATA).

Video Ports

Monitors/computer screens and projectors connect to your PC or laptop using a variety of different connectors.

Many tablets and smaller computers have micro versions of these ports, eg micro USB or micro HDMI

DVI

Digital Video Interface is a video display interface used to connect a video source (eg your computer) to a display device, such as an HD ready TV, computer monitor or projector.

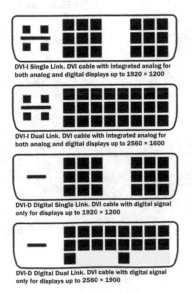

DVI can get a bit confusing, as there are a number of different connectors. Here is a summary.

DVI-I Single Link. DVI cable with integrated analog for both analog and digital displays up to 1920 × 1200

DVI-I Dual Link. DVI cable with integrated analog for both analog and digital displays up to 2560 × 1600

DVI-D Digital Single Link. DVI cable with digital signal only for displays up to 1920 × 1200

DVI-D Digital Dual Link. DVI cable with digital signal only for displays up to 2560 × 1900

HDMI

High Definition Media Interface, is a combined audio/video interface for carrying video and audio data from a High Definition device such as a games console or computer to a high end computer monitor, video projector, or High Definition digital television.

Pictured below is Standard HDMI & Micro HDMI.

VGA

Video Graphics Array is a 15-pin connector found on many computers and laptops and is used to connect to projectors, computer monitors and LCD television sets.

Component Video

Carries a video signal (no audio) that has been split into three component channels: red, green, blue. It is often used to connect high end dvd players to televisions.

Composite Video

Carries an analogue standard definition video signal combining red, gree, blue channels (with no audio) and is used in old games consoles or analogue video cameras.

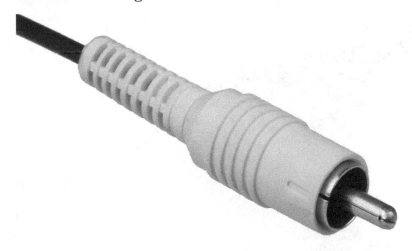

Audio Ports

1/8" (3.5mm) Phono Jack

The phono jack also known as an audio jack, headphone jack or jack plug, is commonly used to connect speakers or headphones to a computer, laptop, tablet or MP3 player and carries analogue audio signals.

1/4" (6.35mm) Phono Jack

These are generally used on a wide range of professional audio equipment. 6.35 mm (1/4 in) plugs are common on audio recorders, musical instruments such as guitars and amps.

3 Pin XLR

The XLR connector is usually found on professional audio, video, and stage lighting equipment.

Many audio mixing desks have XLR connectors to connect stage mics and instruments.

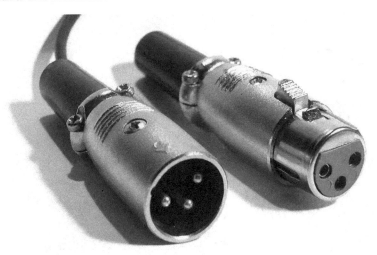

RCA Audio

Some home audio system, televisions and DVD players use RCA cables to connect to audio receivers, amplifiers and speakers.

Types of Computer

There are many different types of computer and they range from the smallest smart phone or tablet computer to large super computers that fill entire buildings.

The most common ones you'll find are micro computers, built on the micro processor.

Micro Computer

A microcomputer is a small, relatively inexpensive computer with a microprocessor that has become commonly known as a PC or Personal Computer. PCs now days come in various different incarnations depending on their function; desktops for power, larger hard disc, memory and larger screens or laptops and tablets for their portability.

Desktop

The traditional desktop computer with a monitor, computer case, keyboard and mouse. Can either be a Mac or a PC.

These machines are usually quite big and have the most computing power. They are aimed at gamers, graphic designers, video editors, office users and professional users. They are ideal with large screens, plenty of computing power and storage space.

Desktop computer sales for home users have been steadily declining in favour of laptops and tablet computers.

Desktop computers still seem to have a place in an office environment, however this seems to be slowly changing toward a cloud based environment where data is stored on the cloud and accessed using laptops or tablets.

Desktop: All-in-one

This type of desktop is virtually identical to the traditional desktop we talked about above, except the computer case has been done away with.

Instead, all the internal hardware (processor, RAM, hard disc and video card) from the computer case, is integrated into the back of the screen itself.

This makes the whole system easier to set up, as all you need to do is plug in your keyboard and mouse, hook it up to the power and you're ready to go.

Some of these systems have touch screens built in, allowing you to tap icons on the screen instead of using a mouse.

Apple's iMac was the first to use this format, but many other manufacturers have copied this design.

Laptop

A typical laptop computer, also sometimes called a notebook. This one is a laptop running Windows 10.

Laptops usually have a similar spec to their desktop counterparts, however there are some compromises due to space. They tend to have less RAM and run slightly slower than desktops. The screens are usually between 12" and 17".

They can run all the software and apps that are available on a desktop and come with Windows 10 or Mac OS.

The major advantage of a laptop, it its portability. The fact that you can use it in any room, sit on the sofa and surf the web, talk to your friends. Or do some college work in a coffee shop or library.

With laptops, you can plug in various peripherals such as a mouse as well as an external screen or projector. This makes them ideal for those who do public speaking, teaching/lecturing, and presentation.

Some laptops nowadays even include touch screens where you can navigate around the screen by tapping icons and menus rather than using a mouse or trackpad.

Netbook

Netbooks are small cut down versions of laptops. They have less RAM, HDD space and are designed to be small, lightweight and inexpensive which makes them great for carrying around.

The screens are usually about 10". Notice the size compared to the ball point pen in the photograph.

Netbooks can run Windows 10, some form of Linux or even Chrome OS.

These are great for working on the go or travelling around. They can run traditional software such as Microsoft Office and work well when browsing the web, social media or keeping in touch via email.

These have limited power, so anything more processor intensive such as Creative Suite or some types of games will struggle to run on these machines.

These machines also have limited storage space, so if you have a lot of music, documents, videos, or photographs, you'll quite quickly run out of space.

Most of these machines can be used with some kind of cloud storage such as OneDrive or GoogleDrive.

Chromebook

A ChromeBook is a laptop or tablet that runs an operating system called Chrome OS and uses Google's Chrome Web Browser to run web apps.

At its core, Chome OS is a linux based operating system and will run on hardware with either intel/amd x86/64 or ARM processors.

ChromeBooks are designed to be used online, meaning you must be connected to the internet all the time whether you are at home, the office, in school, college, the library, or generally out and about. Without an internet connection, your ChromeBook can still function but will be limited at best.

Traditional software such as Microsoft Office, Adobe Creative Suite and many types of games do not run on these machines. However, Google have developed their own alternatives. Instead of Microsoft Office, you'd use Google Docs.

You can also download countless apps from the Google Play Store for all your other software needs from social media and communication, to getting your work done.

Tablet Computers

Tend to be a cut down compact version designed with touch screens. This one is running Windows 10 in desktop mode.

Examples of these come in the form of iPad, Microsoft Connect Tablets, Surface Tablets, Samsung Galaxy Tab, Amazon Fire and many more.

These are ideal for travelling and carrying about as they are light weight and can be stored in your bag easily.

They have countless apps available from the app store that you can download directly onto your tablet. These range from games to cut down versions of Microsoft Office and basic graphics packages. They are also good for browsing the web, social media, making video calls and keeping in touch using email.

Some tablets can even run traditional software, if they are running Windows 10.

Hybrids

Hybrids are a cross between laptop computers and tablets. An example of a hybrid is Microsoft's surface tablet.

These can function as a laptop and have detachable keyboards. Once you detach the keyboard you can use the device in tablet mode, attach the keyboard and you can use it as a laptop.

These devices aren't usually as powerful as traditional laptops and are usually smaller and light weight.

They also have countless apps available from the app store that you can download directly onto your hybrid. These range from games to cut down versions of Microsoft Office and basic graphics packages. They are also good for browsing the web, social media, making video calls and keeping in touch using email.

Some hybrids can even run traditional software, if they run Windows 10.

Mainframe Computer

Mainframe computers operate at a very high speed and have many peripheral devices attached to them. They occupy very large space more often than not an entire room.

Mainframe computers have a large memory capacity and processing power and are used for critical applications such as bulk data processing of census records, statistics, and transaction processing in banks. These computers can have thousands of workstations or terminals logged in at the same time.

Super Computer

This supercomputer shown below is Blue Gene and has over 250,000 processors and can perform an incredible 200 trillion operations per second. The entire computer is grouped into 72 cabinets, each connected by a high-speed optical network and is housed in an air conditioned room.

Supercomputers play an important role in the field of computer science, and are used for a wide range of processor intensive tasks such as quantum mechanics, weather forecasting and climate research where very large amounts of data need to be processed and calculated very quickly.

Embedded System

An embedded system is a small computer designed for a specific purpose and is usually embedded into the device it controls. The computer program is usually stored either in ROM or some kind of FLASH memory and is known as firmware. These systems can be embedded into digital cameras, smart phones, cars, and household devices such as DVD players, washing machines, microwaves and smart TVs.

Chapter 5

Understanding Hardware Specifications

Computer specifications can often be confusing as they tend to include a lot of promotional and technical words and phrases.

In this section we'll go through the most common information you'll most likely need when shopping for a computer, laptop or tablet.

We'll take a look at some common specifications and try to decode all the technical jargon, so you can be confident when shopping around for a computer.

So Many Machines

A lot of people ask what computer I should get. This is a valid question as there are so many to choose from. You could have a tablet, laptop or netbook/chromebook...

...or perhaps a desktop, macbook or iMac.

This is what I have found from experience of using computers at different levels for different tasks. The question I always ask myself is what I will be using the machine for.

- Will I be playing games?
- Will I be editing photos or video? Or just looking at them?
- Will I be typing documents, making spreadsheets or presentations?
- Will I be using the internet, email and a bit of video chat?
- Do I need portability?

You should take all these things into consideration when buying your machine.

For example, if you are just using your machine to type some documents, check your email and browse the internet, you perhaps don't need the most powerful machine you can find as a lot of that power is expensive and could be wasted.

On the other hand if you do a lot of video editing or photography or play the latest games you would need as much power as you can afford.

Do you want to be sitting at a desk while you use a computer? Or do you prefer sitting in your favourite chair in a coffee shop? If you prefer the coffee shop setting or moving around, perhaps consider a laptop.

Mac or PC?

There seem to be two main types of computer available in most computer shops. One is an Apple Mac the other is a Windows PC.

One of the biggest deciding factors in buying a Mac or a PC is the price. Macs are much more expensive than PCs, in some cases costing up to twice as much as a comparable PC. This can make a Mac not worthwhile if you only need to do basic computing.

Apple is very selective about what programs and applications it allows to run on its machines so software availability can be an issue. However if the software is available for Macs, it is generally high quality and reliable, but there is less variety.

Macs tend to be a bit more secure than PCs as there are a lot more internet threats that target PCs.

Macs are used in creative industries, so a large amount of creative software is available for Mac, making them a good fit for graphic designers and people who love taking photographs, listening to music and playing movies. They have built-in software and hardware features that play music or movies, edit and store photographs and also have software for internet and email along with office applications for word processing etc.

PCs tend to be used in offices and businesses so they have a wide range of applications available for internet browsing, email and word processing.

PCs are better suited for gaming because of their powerful graphics capabilities and the wide variety of games available for PC.

PCs tend to be better for users who occasionally use their computer or use it for basic tasks such as internet, email, a few photographs, bit of music and some typing as paying the price for a Mac might not be worthwhile.

However if you use your computer a lot, love taking photographs, making videos, watching DVDs, listening to music as well as typing documents, browsing the internet and using email, a Mac might be worth its price.

Decoding the Jargon

In this section, we'll take a look at come common computer and peripheral specifications you might find online or when shopping for a computer.

Computer Specs

It's worth knowing a bit about these terms before going to the computer store. I have tried to filter out most of the tech-babble and circled the information you need.

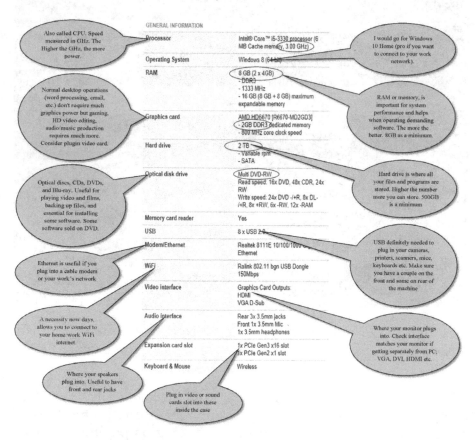

The main components to look for are the processor, graphics card, hard disc, and RAM.

Most modern machines come with a WiFi connection as well as a keyboard and mouse.

Printer Specs

Here is a common specification you might find when shopping for a printer.

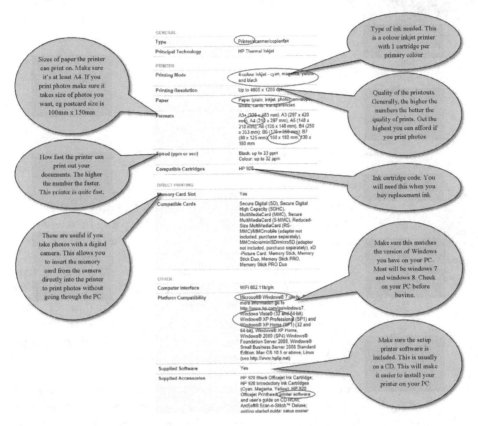

If connecting to your computer using USB, make sure you buy a USB cable, some printers don't include one.

Chapter 6

Computer Software

Computer software comes in a variety of different forms: Applications, Apps, Utilities, and System Software.

Applications are pieces of software that are quite large and were originally designed to run on a desktop computer; you interact with the application using a keyboard and mouse. Example applications are: Microsoft Office Suite, Adobe Creative Suite, and so on.

A more recent incarnation of software is the App. Apps are usually smaller in size and are designed with a touch screen in mind. So for example, facebook has an app. Microsoft Office Suite also has App versions of its software to complement their desktop application counterparts.

Utilities are small programs that are designed to configure, analyse, optimise or maintain a computer, such as antivirus, scandisc or defrag.

The most important piece of software is called the Operating System such as Windows 10, MacOS, iOS or Linux. This is called system software.

The Operating System

The Operating System or OS is a piece of system software that manages all the hardware and software resources available on your computer. This could be memory allocation, storage device management, file management, as well as providing a nice user interface of windows and icons for you to interact with.

On laptops, PCs and some tablets you could have Microsoft Windows 10 as your Operating System.

On a Mac you'd be running MacOS

On an iPad or iPhone you'd be running iOS.

Other tablets will have their own operating system, eg Android.

In each case, the system provides a Graphic User Interface (GUI) for you to work with using menus and icons to represent apps and commands.

User Interfaces

The interface is the way in which you interact with a computer. There are generally two types of user interfaces: The command-line interface and the Graphic User Interface.

Command line interface such as MS-DOS as shown below, allow the user to type in various commands to execute programs and other tasks.

```
C:\Users>dir
 Volume in drive C is System
 Volume Serial Number is 7CF5-8D97

 Directory of C:\Users

12/03/2019  14:28    <DIR>          .
12/03/2019  14:28    <DIR>          ..
27/04/2019  21:57    <DIR>          Admin
27/04/2019  21:57    <DIR>          kevwi
16/03/2019  16:39    <DIR>          Public
               0 File(s)              0 bytes
               5 Dir(s)  155,587,416,064 bytes free

C:\Users>_
```

A graphic user interface on the other hand, displays a graphic environment consisting of icons, windows, and menus.

You interact with this environment using a keyboard and mouse or sometimes a touch screen. The graphic user interface used in Windows and MacOS is the modern approach to interacting with a computer.

Building on the graphic user interface model, many modern devices such as tablets and phones offer a touch screen graphic user interface, sometimes called a Mobile User Interface.

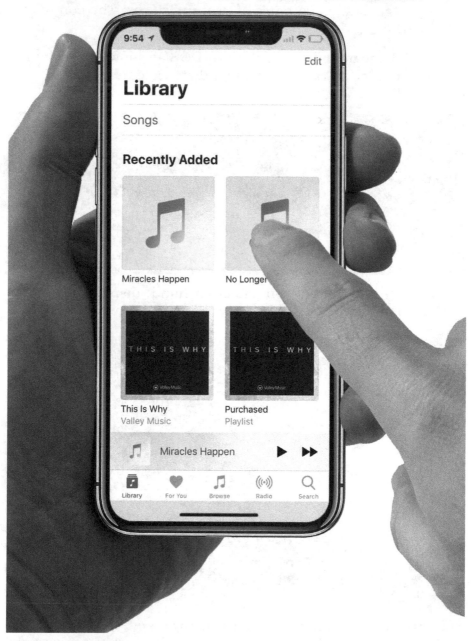

This allows the user to interact with the system by tapping on icons rather than using a mouse.

Process Management

A scheduler is used to allocate processor time to the different processes waiting to be executed. A process is simply a computer program that is being executed.

Multi programming is the ability of an Operating System to execute more than one process at the same time on a single processor, and requires that the processor be allocated to each process for a set period of time.

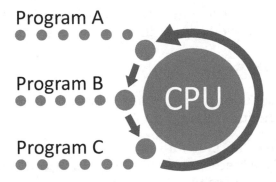

Multiprocessing is the execution of processes by more than one processor.

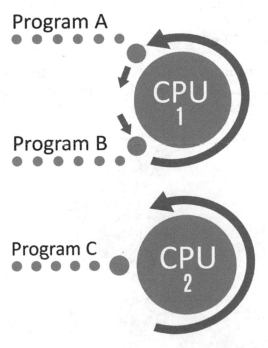

Non Pre-emptive Scheduling

Non pre-emptive scheduling is one in which a process holds the CPU until it is terminated or reaches a wait state and cannot be interrupted.

The advantage is, this type of algorithm is simpler and easier to implement. The problem is short processes have to wait for the long process at the front to finish, making this algorithm less efficient.

Examples of non pre-emptive scheduling algorithms include first come first served scheduling.

Pre-emptive Scheduling

Pre-emptive scheduling is one in which a process is allocated to the CPU for the limited amount of time before being interrupted.

The advantage is scheduling is more robust, all running processes make use of the CPU equally. However low priority processes are susceptible to starvation and long wait times.

Examples of pre-emptive scheduling algorithms include round robin scheduling and priority based scheduling.

First Come, First Served

With this algorithm, the processes are executed in the order in which they arrive. This is very slow since processes have to wait for the previous process to finish before they can begin.

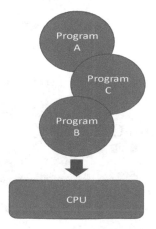

Round Robin

Each process is given a fixed amount of execution time. A process is executed until its time expires. This process is then suspended, and the next process begins its execution...

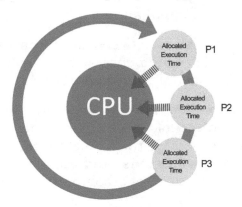

When all the processes in the queue have been allocated an amount of time, the scheduler returns to the beginning of the process queue and starts again. This loop continues billions of times a second, giving the impression processes are running at the same time.

Multilevel Feedback Queueing

Processes enter on the high priority queue. If a process uses too much CPU time, it is moved down to a lower-priority queue.

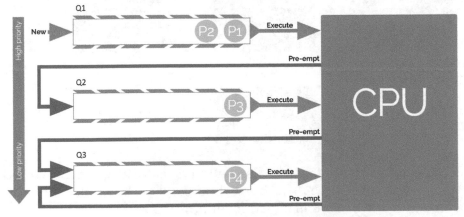

Processes on the lower queues are not serviced until all processes in the queues above are empty. The bottom queue uses a round robin scheduling scheme.

File Management

An operating system allows you to manage files. For example, Windows saves data in the form of a file. This file could be a photograph, eg, claire.jpg; a document, eg lettertoclaire.doc; a music track, eg demotrack.mp3, and so on.

Notice Windows distinguishes between file types using a 3 or 4 letter extension: doc for documents, jpg for photos, mp3 for music etc. This helps to identify different types of data.

These files are stored in a hierarchical structure. The Operating System stamps each file and folder with some data: date created or modified, date last accessed, user ID of owner or creator of the file, various access permissions, and the file size. Here is a common file/directory structure of a computer system.

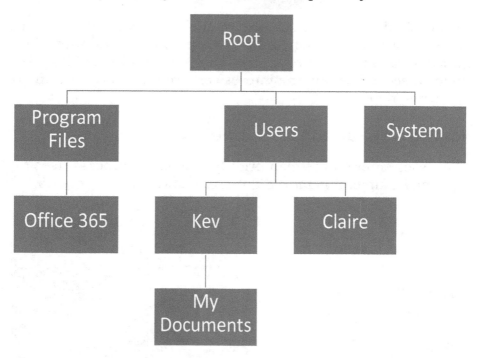

Files are saved into folders (also called directories) and are stored on a hard disc drive. The hard disc drive is located inside the computer itself. This is called local storage.

Data can also be stored on OneDrive or GoogleDrive. This is called cloud storage.

You can perform several functions on files, for example:
- create a file
- rename a file
- save a file to a location
- copy a file to a location
- move a file to a location
- delete a file

The operating system itself can be stored in a directory (eg windows), application software can be stored in another (eg program files).

You can view all the files on your computer using Windows File Explorer. Click the yellow icon on your taskbar. This is where you can organise files, delete files, copy files as well as view files.

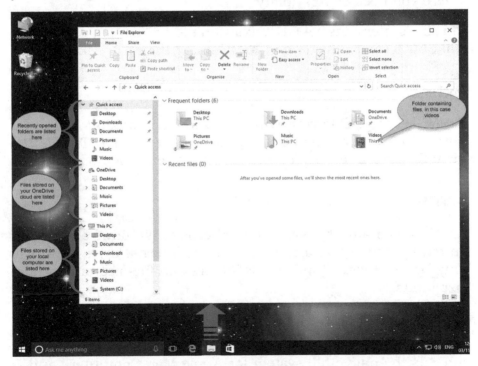

To keep files organised, Windows allows you to create folders to keep all your different types of files grouped together.

These folders (also called directories) can be referenced using a path. For example, the 'my documents' folder:

```
C:\users\kev\my documents\
```

File Systems

A file system manages how and where data is stored on a storage device such as a hard drive. Each operating system has its own file system. Windows uses NTFS, Macs use APFS or HFS.

File systems have directories (also called folders) that allow the operating system to organise files.

FAT32

FAT32 is an old file system that works with all versions of Windows, and Mac, as well as Linux and Game Consoles. This makes it ideal for use on memory sticks and external drives where you need compatibility.

This file system has a 4 GB maximum file size limit and Windows will only format drives up to 32GB using the FAT32 file system.

exFAT

exFAT is compatible with Windows and Mac, making it ideal for portability with no file-size or partition-size limits. This is the ideal file system to use for external drives and memory sticks.

NTFS

NTFS is the native file system for Microsoft Windows and is best suited for your system drive and other internal drives that will only be used with Windows. NTFS has no file-size or partition size limits.

APFS

APFS is the native file system for Apple Mac Computers and is best suited to Apple Macs. APFS has no file-size or partition size limits.

EXT2, 3, and 4

Extended File system, linux machines use this file structure.

Memory Management

A modern computer will have numerous programs running at the same time, so it's vital that the operating system prevents programs from corrupting other programs or data also stored in memory.

There are various memory management schemes that could be employed, such as swapping, segmenting, partitioning, and paging.

The memory management scheme allocates each running program its own memory and keeps track of every memory location. Allocated memory that is no longer required by a program is released and reused.

Memory Partitioning

There are two types: fixed partitioning and variable partitioning.

With fixed partitioning, main memory is divided into a set of fixed non overlapping regions called partitions. Memory partitions don't need to be the same size. In the example below, you have an 8MB partition for the OS, two 16MB partitions for programs.

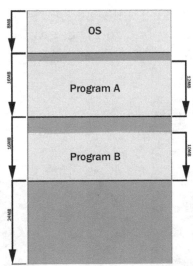

Each program is assigned to the smallest partition it will fit in. Notice not all programs will fit exactly into the partition, so there is fragmentation and wasted space.

With variable partitioning, partitions can be variable in size where each process is allocated exactly as much memory as it requires.

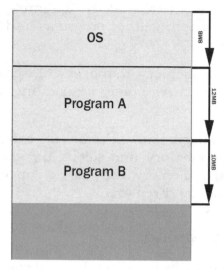

Compaction is used to clean up memory that has become fragmented by shifting processes so they are contiguous, and all free memory is in one block.

Segmentation

Each program is divided into blocks of various sizes called segments which correspond to logical divisions in the program defined by the programmer.

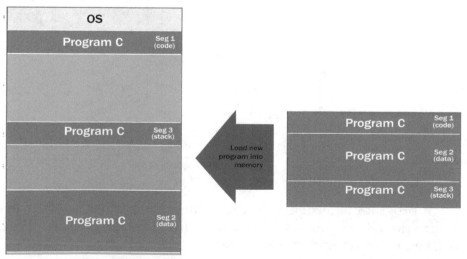

Paging

With this approach, a program's address space is divided up into small 4 KB units called pages - effectively dividing up the program.

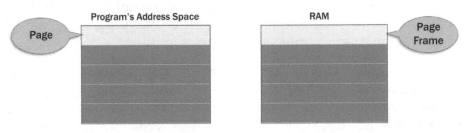

The physical memory (the RAM) is divided up into page frames of the same size.

The executing program's pages are loaded into the next available page frames in physical memory. This helps to make optimum use of the available physical memory without wasting space.

Memory (RAM)

00	Program A Page 1
01	Program B Page 1
02	Program B Page 2
03	Program A Page 2
04	Program B Page 3
05	
06	
07	

Load new program into memory

Program C   Page 1
Program C   Page 2
Program C   Page 3

Page Frame

Memory used by programs that have terminated is released and made available to new programs.

Virtual Memory

Virtual Memory is a storage abstraction combining physical RAM and the hard disc creating one large virtual address space for programs to use. In other words, the operating system uses a portion of the hard disc drive as if it were part of the computer's physical memory (or RAM). This means that programs can be much larger than the physical memory installed.

The problem is, the hard disc drive is much slower than the physical memory (RAM) and has an adverse affect on system performance, since the processor has to wait for data to arrive from the hard disc drive. This is why increasing the physical memory (RAM) increases performance and reduces the need for virtual memory.

Now as with the paging approach, the program's address space (or virtual address space) is divided up into small 4 KB units called pages and the physical memory (the RAM) is divided up into page frames of the same size.

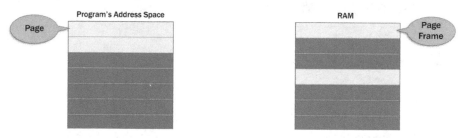

As a program executes, the pages containing the instructions required are loaded into page frames in physical memory.

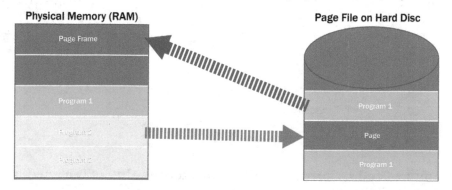

Pages containing parts of the program that are not currently being executed are transferred to the hard disc.

116

The operating system uses a page table to map the virtual addresses used by the program to the physical addresses (or real addresses) used by the hardware.

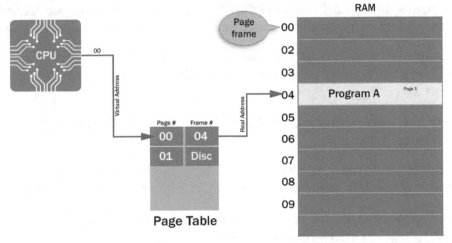

A page fault occurs when a program attempts to access a page that is not currently stored in physical memory. The operating system must then locate the page on the hard disc, load it into a page frame in physical memory, and update the page table.

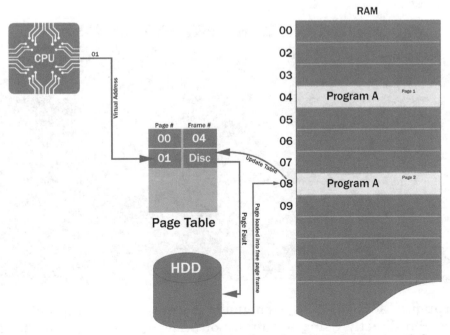

Pages stored in physical memory that haven't been used for a while, are moved to the hard disc.

Apps & Applications

As mentioned earlier, applications are pieces of software that can be quite large and were originally designed to run on a desktop computer; you use the application using a keyboard and mouse.

Examples applications are Microsoft Office Suite: Word, Excel, PowerPoint, Adobe Creative Suite: Photoshop, Adobe Premiere and so on.

On the setup above, we have Microsoft Word running on a desktop computer.

You interact with the system using a keyboard and mouse, so the application and its interface is designed with this in mind.

This is an example of a desktop application. This application could also be running on a laptop.

A more recent incarnation of software is the App. Apps are usually smaller in size and are designed with a touch screen in mind. In the demo below, we have a maps app running on a tablet.

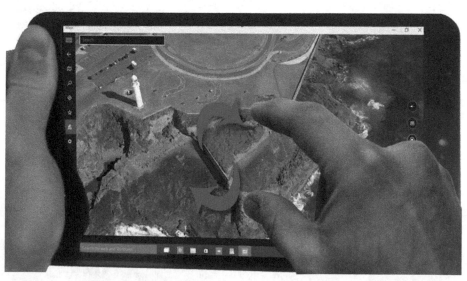

You interact with the system using your finger to manipulate the screen directly using a number of finger gestures; point, drag, tap etc. The interface is designed with this in mind, making icons bigger to enable you to tap on them with your finger.

Anti-Virus Software

A lot of this software is sold pre-installed on the machine you buy and is offered on a subscription basis. So you have to pay to update the software.

There are some that are available for free to home users, such as:

- Windows Defender
- AVG
- AVAST

Windows Defender

Windows 10 comes pre-installed with Windows Defender which is automatically updated by Microsoft for free.

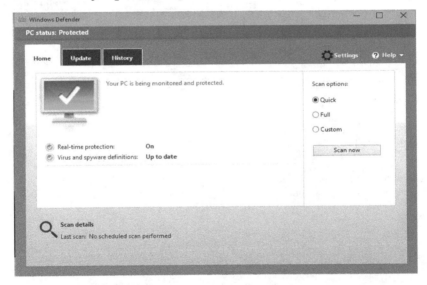

This is the bare essentials and is the minimum protection against viruses and online threats. This is adequate if you just browse the web and check your email. If you do online banking or shop online, then you should have a look at some of the more advanced security software packages.

Two free ones that are a good place to start are Avast and AVG. Both of these packages are very good. The free one is basic, but you can upgrade if you need something more.

Avast

Avast scans and detects vulnerabilities in your home network, checks for program updates, scans files as you open them, emails as they come in and fixes PC performance issues.

You can download it from their website.

www.avast.com

Scroll down the page until you find 'free download'.

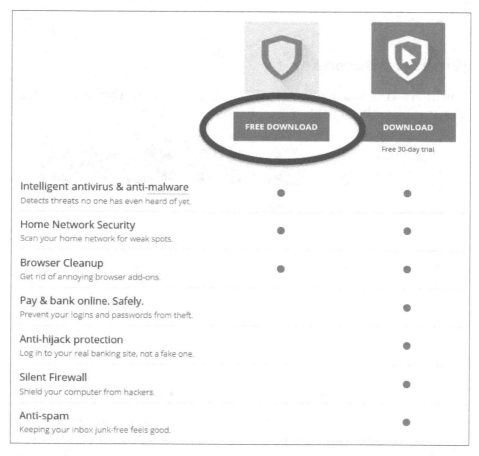

The other version here is a 30 day trial and will expire after 30 days. You will need to pay a subscription to continue.

When prompted hit 'install'. If the installation doesn't run automatically, go to your downloads folder and run 'avast_free_antivirus_setup.exe', follow the on screen wizard.

AVG

AVG blocks viruses, spyware, & other malware, scans web, twitter, & facebook links and warns you of malicious attachments.

You can download it from their website.

www.avg.com

Scroll down and click 'free download' at the bottom of the page.

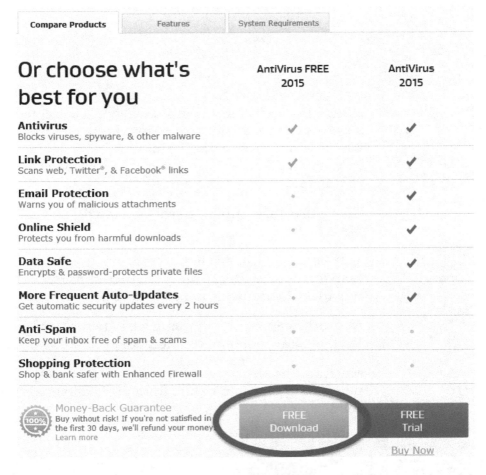

The other version here is a 30 day trial and will expire after 30 days. You will need to pay a subscription to continue.

When prompted hit 'install'. If the installation doesn't run automatically, go to your downloads folder and run the setup exe file.

Computer
Networks

When two or more computers are connected together they are said to be networked. A network can be two or three computers connected together in a small office or connect computers together in different parts of the country.

There are three main types of networks. LAN, MAN & WAN.

You can have either a client server or a peer to peer network model, depending on the environment in which the network is being used.

Networks can be set up using various topologies, such as star, bus, or ring networks.

Machines on the network can be linked together using Cat5 ethernet cable, fibre optic cable, and cellular or wifi.

Local Area Networks (LANs)

A small network contained in a single site or building is called a LAN or Local Area Network.

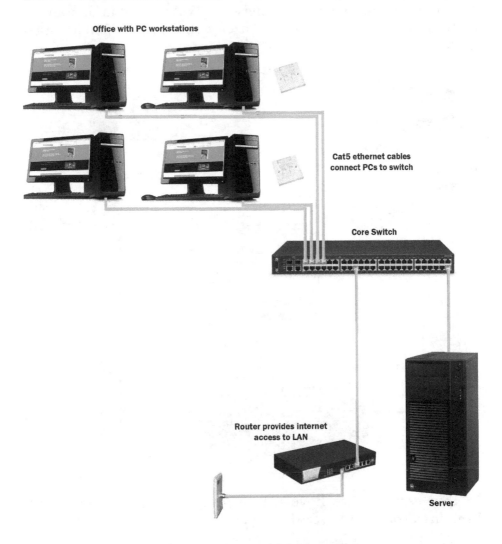

Office with PC workstations

Cat5 ethernet cables connect PCs to switch

Core Switch

Router provides internet access to LAN

Server

As you can see in the diagram, the network covers a small area. The computers could be split up into different offices or all in one room and can all access resources served from the file server and use internet services provided by the router.

They are all connected together using a switch.

Metropolitan Area Networks (MANs)

A metropolitan area network is a network that connects computers in a region larger than a local area network (LAN) but smaller than a wide area network (WAN), and is usually an interconnection of networks in a city that form a single larger network.

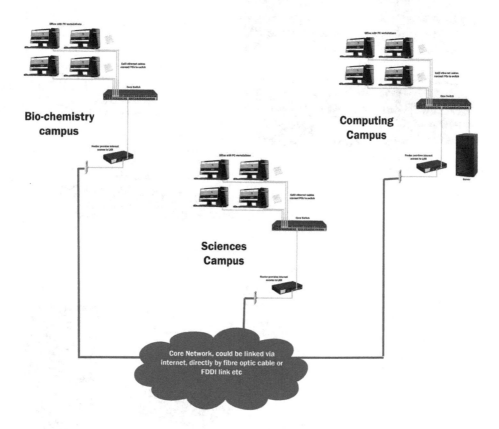

In this example, we have a university campus network with buildings spread all over the city.

Each campus has its own LAN, and the LAN at each campus is linked together to form a much larger network called a MAN.

These could be linked using fibre optic cables or even a virtual private network using the internet.

Wide Area Networks (WANs)

A network that connects computers and LANs together in different parts of the country or world is called a WAN or Wide Area Network.

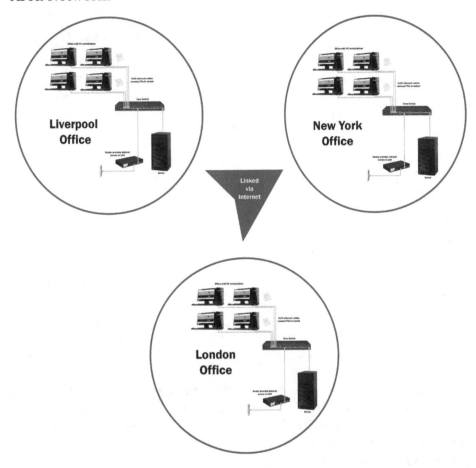

This example is of a multinational organisation that has offices in different cities and countries.

Each organisation can have its own LAN and is linked to a larger network over the internet.

A WAN can also have MANs connected to it as well as lots of smaller LANs.

Peer-to-peer Network

On peer-to-peer networks, all the computers on the network are equal in role, and are usually connected together in a small office or maybe at home.

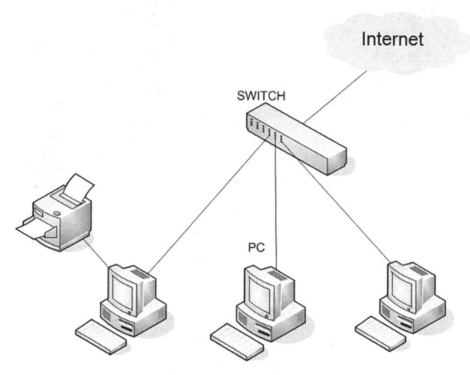

An example of a peer-to-peer network is one that might be found in a modern household, a computer in the study, one in the living room and one in the bedroom, all sharing one internet connection and one printer. Each computer, laptop, tablet or phone is called a peer.

Practically these computers are probably connected using WiFi to a switch that is built into a router provided by an ISP. They could also be connected using ethernet cables to a switch.

Each machine on the network can access the internet, and can also share files and any peripherals connected to the machine, such as a printer.

PC-1 could share a folder containing files which can be accessible from the other PCs, laptops or devices.

Client-server Network

Client-server networks are found in businesses, colleges and places you would find a lot of computers.

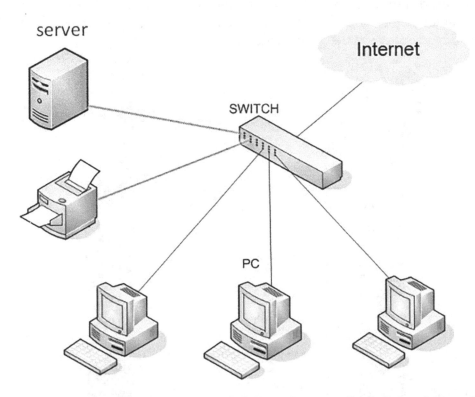

These networks consist of a number of PCs called clients located in offices on an employee's desk, in classrooms and are usually connected using CAT5 cables.

Also on the network there would be one or more servers. Servers are large computers that hold data and shared resources that are served to the client PCs on the network. Hence the name client-server.

The server could contain shared folders containing documents that can be accessed by all the computers on the network.

Similarly the printer would be a print server that each computer on the network has access to and can send documents to print.

Each machine can access the internet.

Network Layers

The OSI 7 Layer Model is a conceptual framework used to describe the functions of a computer network and is split into seven different layers: Physical, Data Link, Network, Transport, Session, Presentation, and Application. Each layer handles a different part of the communication.

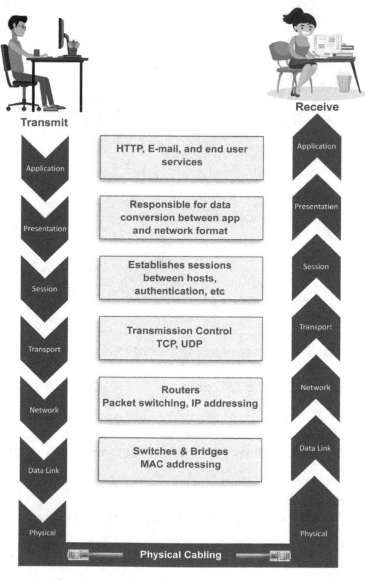

A layer serves the layer above it and is served by the layer below it.

Network Topologies

There are various topologies used to construct networks. We'll take a look at some of the common ones that have been used in the past.

Star

In local area networks with a star topology, each node is connected to a central hub, called a switch, with a point-to-point connection.

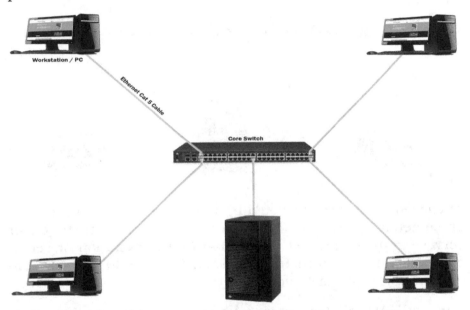

Star topologies are common on Fast Ethernet LANs where each workstation or server connects to a central switch like the one below.

An advantage of the star topology is the simplicity of adding additional nodes. The primary disadvantage of the star topology is that the switch represents a single point of failure.

Bus

In bus topologies although somewhat obsolete in favour of Fast Ethernet now days, each node is connected to a single cable called a bus.

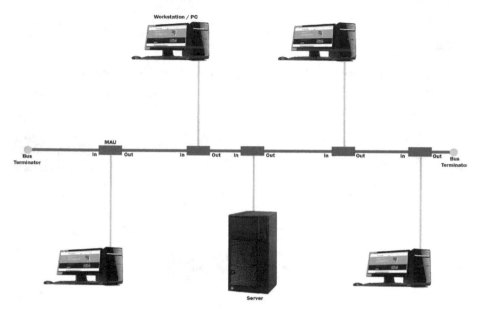

At each node on the bus you'd have a media access unit (MAU), or sometimes simply a T-Connector, where a workstation or server can be connected. More than one workstation or server can be connected to each MAU. Each MAU could be located in different locations or buildings.

The last MAU will also act as a terminator for the bus, meaning the out port on the MAU will have a terminator plugged into it rather than another MAU.

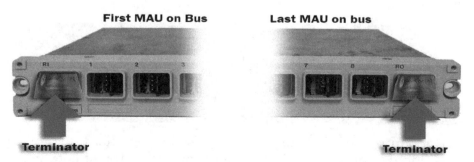

Similarly with the first MAU on the bus.

Ring

Another technology that is somewhat obsolete by today's standards, a ring topology is a bus topology in a closed loop. A token ring network is a common implementation, where a special packet called a token is "passed" around the ring - only the PC with the token is allowed to send data onto the ring. Have a look at the diagram below.

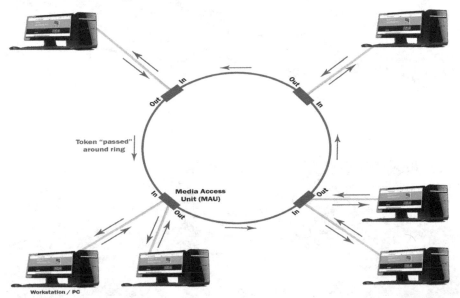

Each node on the ring has a Media Access Unit which looks something like this...

The two ports either end are for the ring coming in, and the ring going out as can be seen on the diagram at the top of the page. The ring could be coming in from another building and going out to another, while all the workstations in the building connect to the MAU.

Ethernet

Computers are usually linked up using copper RJ45/CAT5. This one below is called an unshielded twisted pair (UTP).

...or fibre optic cables

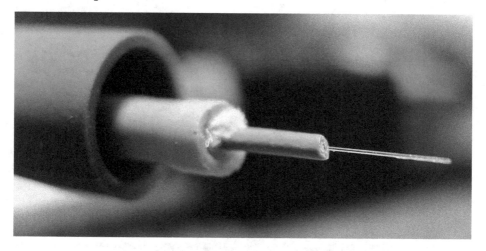

Computers are usually plugged into a device called a switch, shown below.

This device passes data between the different computers on the network.

CAT5 cables connect computers to the switch using either fast ethernet (100BaseT) or gigabit ethernet (1000BaseT), and has a limit of about 100m. The cables are terminated with a standard RJ45 connector.

Note there are 4 pairs of wires (orange, green, blue, brown) and 8 pins. The connectors are wired as shown below.

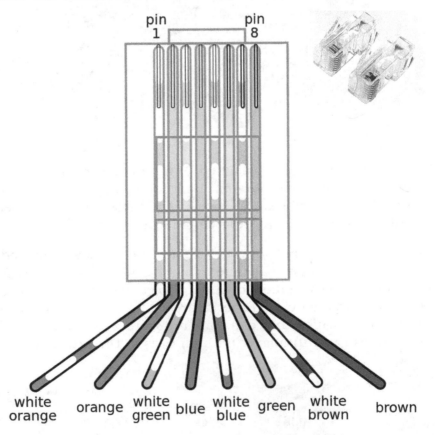

| pin 1 | | | | | | | pin 8 |
| white orange | orange | white green | blue | white blue | green | white brown | brown |

Fibre optic cables are generally used when distances are far greater and can link different sites together.

Cellular Networks

A cellular network or mobile network is a wireless network distributed over land that is divided into areas called cells, such as GSM, 2G etc.

Each of these cells is assigned a frequency (eg F1-F4) each served by a radio base station. The frequencies can be reused in other cells, provided that the same frequencies are not reused in neighbouring cells as it would cause interference.

On a GSM cellular network, frequency in each cell is divided up using a process known as time-division multiple access (TDMA). This allows large numbers of phones to connect at the same time using same frequency channel in the cell. Each user is allocated a time slot in which they can transmit or receive. When the time slot expires the multiplexer (mux) moves to the next user.

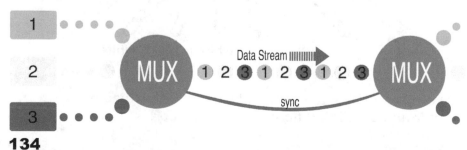

WiFi

WiFi allows you to connect to a wireless network, also called a Wireless LAN and is usually broadcasting on a frequency of 2.4GHz and 5GHz radio bands.

Wireless LANs are usually password protected to keep them private and to prevent unwanted visitors using your WiFi. WiFi networks usually have a network name often called an SSID.

Dual Band or Single Band?

Dual Band wireless LANs use both 2.4GHz and 5GHz but you'll need to make sure your devices (phone, laptop, tablet and computer) are compatible with these frequencies. Some devices only broadcast on 2.4GHz and some use both, so check the WiFi specs on your devices. There is far less interference on the 5GHz band and in some cases can provide better service.

Single Band wireless LANs use either 2.4GHz or 5GHz, not both.

WiFi Extenders

The technical term is wireless repeater and if you live in a big house, these can help to cover your whole property if your WiFi router doesn't quite reach.

The idea is to position the extenders as far away from your wireless router as possible without losing too much of your signal quality. This will give you maximum range.

Wireless Standards

All wireless networks are based on the IEEE 802.11 standard.

The 802.11b standard, has a maximum raw data rate of 11 Mbps using 2.4GHz and is an out dated technology now days.

The 802.11g standard, also known as Wireless G, extended the throughput to up to 54 Mbps using the same 2.4 GHz band.

The 802.11n standard, also known as Wireless N, extended throughput over the two previous standards with a significant increase in the maximum data rate from 54 Mbps to 300 Mbps, and can be used on the 2.4 GHz or 5 GHz frequency bands.

The 802.11ac standard, broadcasts on the 5GHz band and has throughput of up to 1 Gbps and is sometimes referred to as Gigabit WiFi. This is accomplished by using wider RF bands for each channel.

IEEE Standard	Frequency	Speed	Transmission Range
802.11	2.4GHz	1 to 2Mbps	Up to 20 meters indoors but range can be affected by walls.
802.11a	5GHz	Up to 54Mbps	Up to 30 meters indoors but range can be affected by walls.
802.11b	2.4GHz	Up to 11Mbps	Up to 50 meters indoors but range can be affected by walls.
802.11g	2.4GHz	Up to 54Mbps	Up to 50 meters indoors but range can be affected by walls.
802.11n	2.4GHz/5GHz	Up to 600Mbps	Up to 70 meters indoors but range can be affected by walls.
802.11ac	5GHz	Up to 1300Mbps	Up to 40 meters indoors but range can be affected by walls.

Wireless Security

There are currently two standards for home WiFi: WPA and WPA2, WPA2 being the more recent standard.

WPA Stands for 'WiFi Protected Access' and is implemented using a pre-shared key (PSK). It is commonly referred to as WPA Personal, and uses the Temporal Key Integrity Protocol (TKIP) for encryption.

WPA2 uses Advanced Encryption Standard (AES) for encryption. The security provided by AES is much more secure than TKIP, so make sure your WiFi router has WPA2-PSK encryption.

Bluetooth

Bluetooth is a wireless technology for exchanging data over short distances using a frequency of 2.4 to 2.485 GHz. This is often referred to as a PAN (or personal areal network) and can be used to connect headphones, wireless mice, smart phones and make small data transfers.

Many smartphones now use Bluetooth to connect different accessories and can even link to your car's stereo system to make a nice hands free kit, allowing you to safely take calls while driving.

Chapter 8

The Internet

The Internet is a global system of interconnected computer networks linked together using the TCP/IP protocol, and evolved from a research project to develop a robust, fault-tolerant communication network back in the 1960s, known as ARPANET.

Today, you can connect to the internet in a variety of different ways. Many of today's Internet Service Providers offer a DSL, Cable or Fibre Optic connection to the internet, depending on where you are.

In this section, we'll take a look at some of the most common methods and their basic setups.

We'll also take a look at some of the WiFi options and basic internet protocols.

Internet Connections

There are 3 main types of internet connection: DSL, Fibre Optic and Cable. There are others, but they are less common.

Let's go through the main options and see how they work.

DSL

Stands for Digital Subscriber Line and is basically implemented as **a**DSL, meaning the download speed is faster than the upload speed. This type of internet connection connects via your telephone line, allowing you to use both your phone and internet at the same time using a DSL filter.

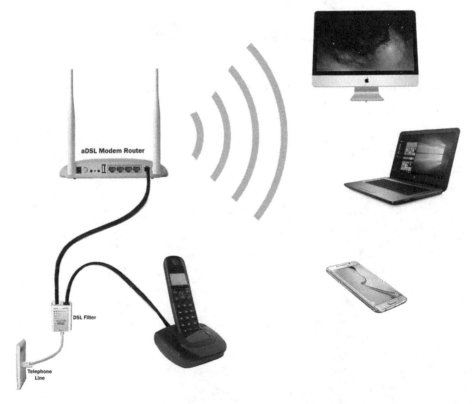

You'll need an **a**DSL / **a**DSL2 modem router that plugs into your phone line. These are usually supplied by your ISP, so check with them before buying.

Fibre Optic

In some countries, the fibre optic cable runs from the exchange to the telephone cabinet in your street and uses **v**DSL over the copper phone line to run the last 100-300m or so to your house.

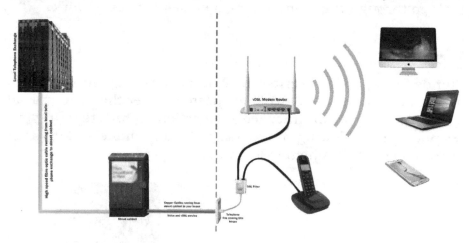

This is called FTTC or 'fibre to the cabinet' and has a very similar setup to the illustration above.

For this option to work you will need a modem router that is compatible with **v**DSL/**v**DSL2. Check with your ISP for specific details.

If you're lucky enough to can get fibre running directly to your home, this is called FTTP or 'fibre to the premises' and is set up as shown below.

This means the fibre optic cable runs from the exchange all the way to your house.

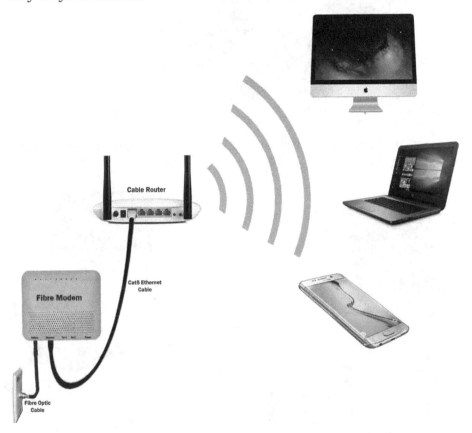

The fibre optic cable will plug into a modem supplied by your ISP which will connect you to the internet.

You can then buy a cable router that has WiFi capabilities and plug that in using an ethernet cable. This will allow you to have WiFi in your house.

Some ISPs will already have this built into their modem, so check with them first.

All your devices such as your laptop, computer, phone, and tablet will connect to the internet through your modem.

Cable

Cable Internet is distributed via your cable TV provider and usually runs down a COAX cable rather than a phone cable.

Setups may vary slightly from different providers, however most will be similar to the one illustrated below.

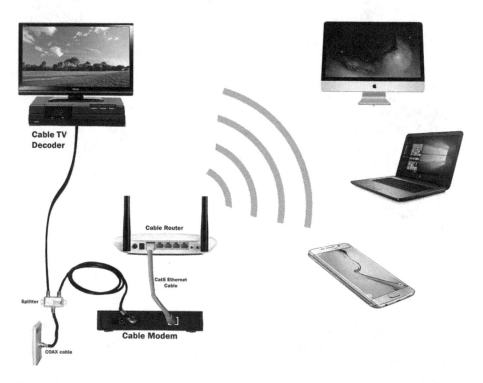

The COAX cable is split and one goes to your cable TV decoder and the other to your cable modem. From your cable modem, you can connect a cable router using an ethernet cable, which can provide WiFi. Some ISPs will already have this built into their modem, so check with them first.

Satellite

This option is available in rural areas where line based broadband services such as DSL or Fibre aren't available.

It uses a satellite dish to provide access but speeds tend to be lower and weather conditions can interfere with reception.

3G/4G/5G

This option uses the mobile/cell phone network and usually involves plugging a USB dongle with a SIM card into your computer.

3G and 4G are usually included with smart phones as part of your package or contract.

Speeds have improved over the years, however they are still very slow in comparison with DSL, Cable or Fibre Optic.

Web Servers and DNS

Web servers are computers usually running Windows Server or more commonly, some flavour of the Linux Operating System such as CentOS. Running on these machines is a piece of software called a web server. This is usually Apache, IIS or NGNIX (Engine X).

In the lab demonstration below, the server on the right is running CentOS Linux and has the Apache Web Server installed. The web server is pointing at our public_html directory stored on one of the server's hard drives.

For simplicity's sake, the Apache web server is bound to port 80, which is the default port for non-encrypted connections.

The IP address of the server is 192.168.0.100.

To access the website on the your computer - the laptop, in your web browser you would need to type in 192.168.0.100. Bit of a pain, just imagine if every website you wanted to go to, you had to remember some string of numbers.

Fortunately we don't have to thanks to DNS servers, which convert our memorable domain names into IP addresses for us.

To keep things simple for this exercise, we wont be setting up a DNS server in the lab, but it's worth remembering the function of a DNS server on the internet.

When you enter the domain name into your browser, your computer will send the domain name to a DNS server. The DNS server responds with the IP address (eg 192.168.0.100).

Your computer (the laptop in the illustration below), uses the IP address to connect to the web server using a port. In this case port 80.

The web server at 192.168.0.100 is listening on port 80, as it was bound to it earlier remember. You can see the configuration summary on the screen in the illustration below.

Once a connection is established, the web server then reads the HTML files in our public_html directory, and then sends the code in the index.html file to your computer - the laptop.

The browser on your computer then reads the HTML code and creates the web page on your computer.

Try it out in the lab.

DHCP Servers

DHCP stands for Dynamic Host Configuration Protocol and is responsible for automatically allocating IP addresses to devices on a network.

Your internet broadband router at home contains a DHCP server to automatically allocate each device an IP address, whether it is your phone, laptop or PC.

Devices connecting to a network broadcast a request to the DHCP server. The DHCP server sends back an IP address from its address pool along with a time period for which the allocation is valid called a lease, as well as DNS and gateway addresses, and the network's subnet mask.

Here's an example of a DHCP server running on Windows Server in our test lab. We have configured the address pool to allocate addresses from 192.168.1.1 to 192.168.1.100.

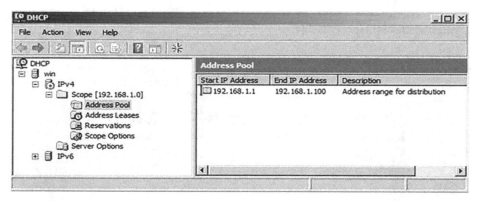

As you can see, one client has connected to the network and has been allocated an IP address.

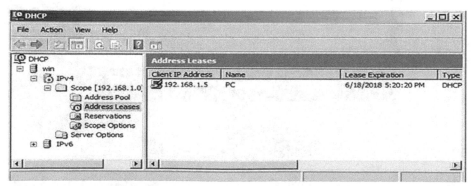

IP Addresses

An IP address identifies each device on a network. There are currently two versions of IP addressing: IPv4 and IPv6.

An IPv4 address is a 32 bit binary number divided into four 8 bit octets, and is usually expressed as a dotted decimal number. For example:

```
89.71.222.37
```

All web servers, email servers, routers, and other devices on the Internet are allocated a unique IP address. These are known as public IP addresses and are assigned by the Internet Assigned Numbers Authority (IANA). Your ISP will assign one of these public addresses to your broadband router. Your router will then allocate a private IP address to each device that connects to your router.

The IANA have allocated the following blocks of IP addresses for private use. Most home internet routers use class C but other larger networks can use either class A or B.

Class A	10.0.0.0 - 10.255.255.255	up to 16,777,216 devices
Class B	172.16.0.0 - 172.31.255.255	up to 1,048,576 devices
Class C	192.168.0.0 - 192.168.255.255	up to 6,5536 devices

If you type ipconfig into your command prompt in Windows, you'll see the current IP address configuration. Notice here, this machine is running on a private LAN and is using Class C private IP addressing.

```
Ethernet adapter Ethernet:

   Connection-specific DNS Suffix  . : lan
   IPv4 Address. . . . . . . . . . . : 192.168.0.133
   Subnet Mask . . . . . . . . . . . : 255.255.255.0
   Default Gateway . . . . . . . . . : 192.168.0.1
```

There are other addresses that have a specific purpose on a network.

- 0.0.0.0 represents the network itself.
- 255.255.255.255 is reserved for network broadcasts, and messages that should go to all devices on the network.
- 127.0.0.1 is called the loopback address, also called localhost meaning a device's way of identifying itself on the network.

- 169.254.0.1 to 169.254.255.254 is assigned automatically when a device can't get an address from a DHCP server.

With IPv4 you can have a maximum of 4,294,967,296 devices. Due to the rapid growth of the internet, this address space has run out. To combat this problem, IPv6 was developed.

IPv6 uses eight 16-bit hexadecimal numbers separated by a colon. Here, we can see in the ipconfig information for this computer, and IPv6 address has been allocated.

```
Ethernet adapter Ethernet:

   Connection-specific DNS Suffix  . : lan
   IPv6 Address. . . . . . . . . . . : fdaa:bbcc:ddee:0:71a4:c9d5:57d4:ec99
   Temporary IPv6 Address. . . . . . : fdaa:bbcc:ddee:0:6165:c608:5408:c037
   Link-local IPv6 Address . . . . . : fe80::71a4:c9d5:57d4:ec49%12
```

An IPv6 address can be broken down into various parts to identify the site or organisation, network, and device. Let's take a look at the address we found using ipconfig.

fdaa:bbcc:ddee:0000:71a4:c9d5:57d4:ec99

Site identifier Subnet Device ID

The first three sets of numbers identify the organisation or site.

fdaa:bbcc:ddee

The next set of numbers identify a subnet or network within the organisation.

0000

The last four sets identify the device on the network.

71a4:c9d5:57d4:ec99

Subnets can be used to divide up large networks into more manageable smaller networks called subnets. These subnets could be departments within an organisation or faculties within a college. Home networks usually only have one subnet as they are small enough to manage.

Subnetting makes managing the network much easier than having one very large network.

TCP/IP

TCP/IP stands for Transmission Control Protocol/Internet Protocol and is a suite of communication protocols devices use to communicate over the internet.

TCP/IP Model

The TCP/IP model, like the OSI model, uses a layered approach and has four layers.

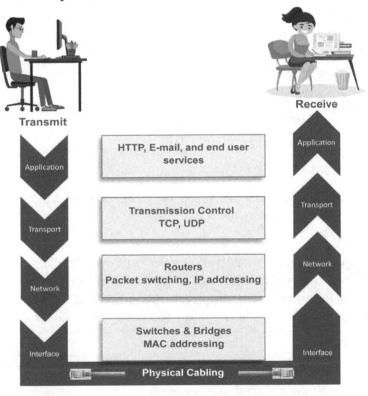

Application Layer deals with network applications such as DNS, FTP, HTTP, IMAP and SMTP. Also browsers & email programs.

Transport Layer utilizes UDP and TCP to establish connections and ensures accurate transmission of data.

Network/Internet Layer uses Internet Protocol (IP) to send packets called datagrams using packet switching.

Network Interface Layer handles the hardware (network cards etc) and physical media such as cables.

Ports and Sockets

A computer on a network may send data to or receive data from multiple computers at the same time. The problem is, these computers have no way of knowing which data belongs to which application. You can identify the application using a port number.

Common application layer protocols have been assigned port numbers in the range of 1 to 1023. Some common ones are:

- HTTP is port 80
- HTTPS is port 433
- IMAP is port 143
- SMTP is port 25

These ports are assigned to specific server services by the Internet Assigned Numbers Authority (IANA).

Port 1024 to 49,151 are ports an organization can register with IANA for a particular service.

Port 49,152 to 65,535 are used by client programs to assign to connections and sessions. These are also known as dynamic or ephemeral ports.

When you connect to a web server, a new TCP connection is established between the server and the client computer. The server can be listening on port 80. The TCP protocol knows which application to send the incoming data to based on the port number received in the data's header.

On the client computer, the connection is assigned a port number between 49,152 and 65,535 so that returning traffic from the server can be identified as belonging to the same connection.

The IP address and the port number form a socket. There will be a socket on the server and one on the client.

We can confirm this on the client using the netstat command as shown below.

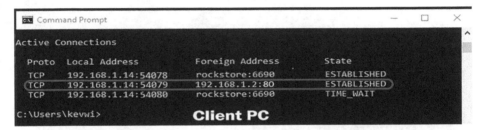

And on the server, we can confirm the connection using netstat as shown below.

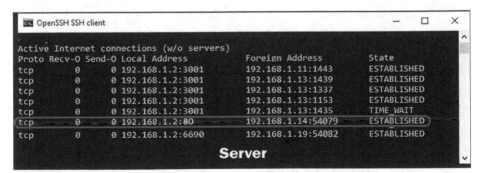

Each socket is unique and bi-directional, so a machine can send data to, and receive data from multiple other machines on the internet.

Try typing `netstat` into the command prompt on a windows machine and you'll see all the current TCP connections.

Packet Switching

On a packet switched network, data is divided into small units called packets.

Remember the Internet is a global system of interconnected computer networks all linked together using routers, switches and cables, in a mesh type system where there are various different routes packets could travel. In the diagram below, each router could connect a different network and could be anywhere in the world. Each network (illustrated by the blue ellipse) could be a college, school, business, or an ISP providing internet access to home users.

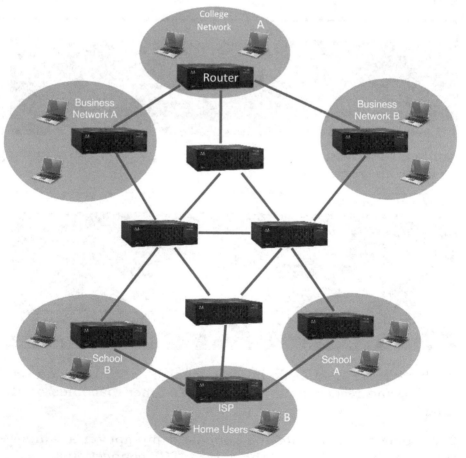

Say we are connected to the internet using a machine on our college network (labelled A at the top of the diagram) and wanted to send an email message to someone.

The email message (or data) is divided up into packets - the number of packets depends on the size of the original data. Each packet includes a header that contains the source & destination IP addresses, fragmentation offset, and a packet ID.

For simplicity sake, lets say the message is divided up into 3 packets (illustrated by the coloured envelopes in the diagram below). Each packet is sent along a different route (illustrated by the coloured lines)

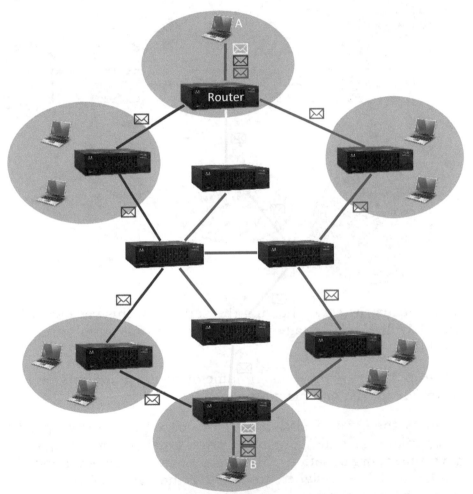

After reaching the destination through various different routes on the network, the packets are arranged in the original order according to the packet ID and fragmentation offset, there by reconstructing the message.

What is a URL?

Each web site on the World Wide Web has an address called a URL or Uniform Resource Location.

The URL itself can be broken down into its basic elements. Lets take a closer look at an example.

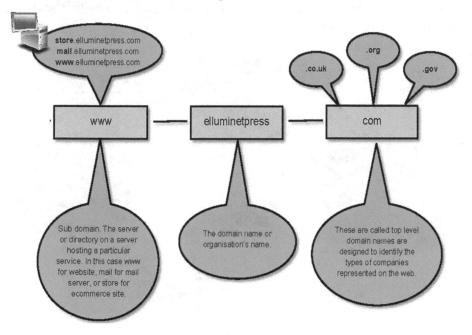

www means the server hosting the service, in this case www for World Wide Web. Usually points to your public_html directory on the web server.

elluminetpress is the domain name or organisation's name and is unique to that organisation.

.com is the type of site. It can be .co.*x* for country specific companies (eg .co.uk), .org for no profit organisations, or .gov for government organisations. These are known as top level domain names and are designed to identify the types of companies represented on the web.

Remember the internet and webservers only understand numbers and IP addresses, so a DNS server will convert the URL domain names into IP addresses which can be routed across the internet.

154

HTML

HTML4 had quite limited functionality and relied heavily on browser plugins such as java, flash and silverlight to play video, audio, games, layout and other functionality. It was a total nightmare to develop websites using HTML4 since it didn't cope well with different screen sizes and different platforms such as tablets and smartphones.

The latest iteration of HTML is version 5. HTML5 brings device independence meaning websites can be developed for all different types of platforms, from PCs to smartphones, without the need to endlessly install plugins on your browser, or develop multiple versions of a website for mobile devices.

HTML5 also introduces some new tags to handle page structure such as <section>, <head>, <nav>, <aside>, <footer>, and some tags to handle media such as <audio>, or <video>.

Where are the HTML Files Stored?

On a web server, the HTML files are all stored in a directory called PUBLIC_HTML.

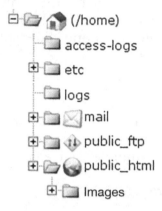

Inside the PUBLIC_HTML folder, we can create sub folders. You can create one for IMAGES and any other resources you want to make available.

Web servers are usually rack mounted and fixed into communications cabinets in server rooms. Each web server is allocated a public IP address which is used by the DNS server to translate the URL typed into a web browser. A single web server might look something like this...

This particular server is running centos linux as its operating system of choice. Some web servers can also run Windows Server.

Cloud Computing

The "cloud" was originally a metaphor for the internet, and many network diagrams represented the internet with a symbol of a cloud.

As internet services advanced, the cloud became a set of hardware devices including data servers and application servers that provide services such as email, apps and disc storage for documents, music and photos.

These cloud services include the delivery of software and remote storage space having as little as possible stored on the user's machines.

If you have an account with a web-based e-mail service like Hotmail or Gmail, then you've had some experience with the cloud. Instead of running an e-mail program on your computer, you log into a web e-mail account remotely.

The storage for your account doesn't exist on your computer, it's on the cloud. In this example, files would be stored on OneDrive and if you're using Google, your files would be stored on GoogleDrive.

You can run Microsoft Word online using your tablet, laptop or phone to edit your documents. If you are using GoogleDrive, you can use GoogleDocs to edit your document. All these can be used through a web browser on your device.

You can also collaborate with other users; colleagues or friends. You can share photos or documents for them to see and edit; working on projects together around the world, or just share the latest photo with a friend.

This has become a huge advantage as data can be stored centrally making backups easier. Applications and servers can be built and maintained centrally by dedicated support staff making downtime a minimum.

Your files on the cloud are stored on a server in a server farm rather than locally on your computer. In the photograph below, there can be about 20 or more servers stacked up in each cabinet and hundreds of cabinets filling entire rooms serving millions of people who subscribe to the service.

Cloud computing can be further broken down into different types

Software-as-a-service or (SaaS): Office 365, Google Docs or web based emails are examples of this where applications are designed for end-users and delivered over the web

Platform-as-a-service or (PaaS): is the set of tools and services designed to make coding and deploying applications quick and efficient

Infrastructure-as-a-service or (IaaS): 'Rackspace' web hosting provider is an example of this and provides the hardware and software that powers the cloud, such as servers, storage, networks, operating systems and so on.

Data Transfer Rates

Say I had an internet connection of 40mbps (megabits per second) and wanted to download a 70mega**byte** file.

First you need to make sure both units are the same. Since the file is measured in mega**bytes**, you need to convert the connection speed of 40mega**bits** per second to mega**bytes** per second.

40 mega**bits** per second ÷ 8 = 5 mega**bytes** per second

So how long will it take to download?

To calculate this use

time to download = file size ÷ data rate

So...

Time to download = 70MB ÷ 5MBps (Mega Bytes per second)
                   = 14 seconds

What about a 5 Giga**byte** file?

Since we are using Mega**bytes** per second to measure the speed of the connection, we need to convert Giga**bytes** to Mega**bytes**.

5GB x 1024 = 5120 Mega**bytes** (remember the units need to be the same; can't use gigabytes and megabytes)

5120MB ÷ 5 Mega**bytes per second = 1024 seconds** (no one quotes that many seconds so you can divide this by 60 to get minutes)

1024 seconds ÷ 60 = approx 17 mins

MBps = Mega Bytes per second

Mbps = Mega Bits per second.

So when your broadband provider boasts "40meg" it is actually 40 mega**bits** per second, not 40 mega**bytes** per second. A lot slower than it sounds, since 40Mbps = 5MBps.

Chapter 9

Internet Security

Internet security is important these days with the rise of cyber crime, phishing, and data theft.

In this chapter we'll take a look at some of the most common threats on the internet today, as well as some social engineering techniques used to trick users into handing over sensitive information.

We'll also take a look at some security and prevention methods that can be used to combat common threats.

Let's begin by taking a look at some of the different types of malware that are around.

Malware

Malware such as worms, viruses, and trojans are designed and written to exploit various different vulnerabilities in computer systems, particularly the operating system and web browsers in order to destroy data, take over a computer, or steal sensitive information.

Virus

A computer virus is a piece of malicious code that replicates itself by injecting code into other programs.

Worm

A worm is a standalone, self replicating, malicious computer program designed to cause disruption to a network, steal data, install a back door, or even lock files in a ransomware attack.

Trojan

The name and concept was derived from the story of the Trojan horse used to invade the city of Troy.

A trojan masquerades as an ordinary program or utility that carries a hidden, more sinister function. This could be data theft, or a ransomware attack.

Rootkit

A rootkit is a program designed to enable unauthorised, remote administrator access to a computer by opening a backdoor.

Rootkits don't replicate themselves like worms and viruses do and are usually installed through phishing scams, executable files, or software downloaded from a dodgy website.

Once installed, an attacker can connect to the infected machine and introduce other malware to steal information. Rootkits usually hide themselves on the infected machine and can be difficult to detect.

Ransomware

Ransomware is a malicious program designed to encrypt and lock a computer system until a fee is paid effectively holding the system to ransom.

Ransomware attacks are usually carried out using a trojan worm that a user is tricked into opening. These trojans can arrive through email or from a phishing scam.

The infamous wannacry worm attacked the British National Health Service (NHS), where hospitals had to turn away patients or cancel scheduled appointments.

Social Engineering

Social engineering is the use of deception and manipulation to trick individuals into handing over confidential information such as passwords, bank details, and personal information in order to steal money or the person's identity.

Phishing

Pronounced 'fishing', this social engineering technique is designed to trick a user into handing over confidential information. Many phishing scams come via an email or phone call, that appear to originate from a legitimate source such as a bank, the police, IRS (HMRC), or a well known company. Here's an example.

The email appears genuine and is designed to scare you, but notice the links have been spoofed.

Following the link will take you to a fake website that looks exactly like the real thing, but notice the URL is fake...

Pharming

Pronounced 'farming', this social engineering technique is designed to redirect a legitimate website's traffic to a duplicate fake site by either using the infected computer's hosts file, the victim's internet router, or a compromised DNS server.

Using the fake site, the scammers can harvest usernames and passwords entered by unsuspecting users, then use them to gain access to the victim's accounts on the official website.

Pretexting

This social engineering technique uses a fabricated scenario designed to gain a user's trust and trick them into handing over personal information.

This scam is usually a phone call that appears to come from an authority such as the police or tax office, but could also impersonate a co-worker, an insurance company, some fake organisation running bogus special offers, or some external IT support company.

Shouldering

Shouldering is a social engineering technique used to harvest passwords, PIN numbers and other sensitive data by discretely looking over someone's shoulder while they enter the information.

A typical scenario is a thief looking over someone's shoulder when using a cash machine, or when entering a pass code to unlock a phone or computer. So watch your back!

Threat Prevention

There are various countermeasures that can be employed to reduce the risk of attack or data theft.

Biometric Measures

New phones, tablets, and laptops nowadays, include finger print scanners, iris, and facial ID cameras, that use a finger print, or a mathematical representation of your face.

Strong Passwords

Instead of using a password like 'password123', you should use a mixture of upper and lower case letters, a number or two, and a couple of symbols such as an underscore, or a dollar sign. The more characters you use, the more difficult the password is to crack. You should also make passwords between 8 and 12 characters long.

2-Factor Authentication

Many online services are starting to include 2-factor authentication where a confirmation code is sent to the user's phone number or email address that was registered when the account was opened.

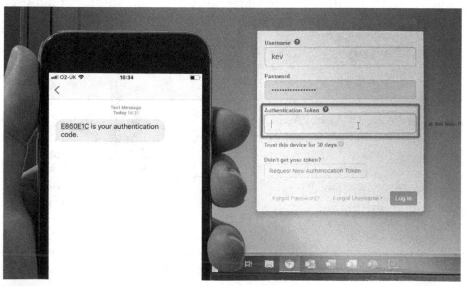

Public Key Cryptography

Public key cryptography, also known as asymmetric cryptography, is an encryption scheme that uses two non-identical keys - a public key and a private key.

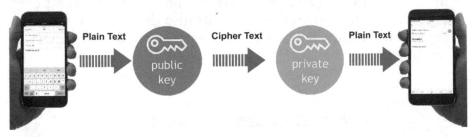

Public keys can be shared freely and are used to encrypt a message. Only users with the private key can decrypt the message.

This scheme is widely used in SSL and HTTPS to establish secure connections and also in the generation of digital signatures and certificates.

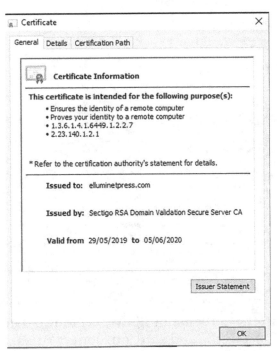

Here is a digital certificate used to secure a website using SSL.

Firewalls

A firewall is a device or software program that monitors network traffic going into and out of a single computer or a network.

A firewall could be a software program installed on a computer. Windows 10 has its own firewall built into the operating system. This monitors data flowing in and out, as well as connection attempts.

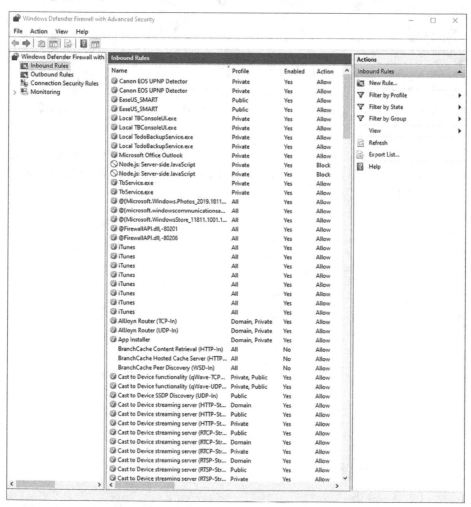

You can see in the screenshot above, windows defender firewall. You'll see a list of inbound and outbound rules. This is where you configure the firewall to allow or block traffic from various services and apps installed on the computer.

A firewall could also be a hardware device. These firewalls are usually found on a network. The firewall itself could be installed on the router or on a separate hardware device.

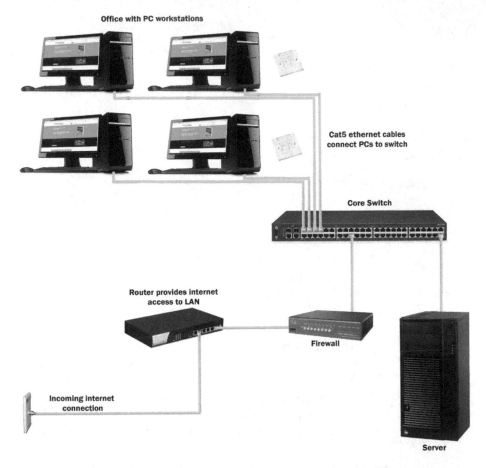

Firewalls employ several techniques to inspect data flowing in and out through the firewall: packet filtering and stateful inspection are two common ones.

Packet-filtering examines each packet that enters the firewall and tests the packet according to a set of rules that have been set up.

Stateful packet inspection looks at packets in groups rather than individually. It keeps track of which packets have passed through the firewall and can detect patterns that indicate unauthorized access.

Index

Index